Seashore
Plants of
California

California Natural History Guides

Arthur C. Smith, General Editor

Advisory Editorial Committee:

California Natural History Guides: 47

E. Yale Dawson
Michael S. Foster

SEASHORE PLANTS OF CALIFORNIA

New Illustrations by Bruce Stewart

UNIVERSITY OF CALIFORNIA PRESS
Berkeley · Los Angeles · London

Figure and Photo Acknowledgments

Illustrations on pages 48, 53, 55, 61, 64, 67, 73, 79, 87, 89, 91, 93, 99, 102, 107, 110, 113, 116, 118, 119, 123, 127, 129, 131, 138, 141, 142, 144, 147, 150, 156 taken from MARINE ALGAE OF CALIFORNIA, by Isabella A. Abbott and George J. Hollenberg, and used with the permission of the publishers, Stanford University Press © 1976 by the Board of Trustees of the Leland Stanford Junior University.

Figures 57a, 57b, 57c, 58c, 58d, 59a, 59c, 61a, 61b, and 63b are from Munz, *Shore Wildflowers of California, Oregon, and Washington*, and *California Spring Wildflowers*, UC Press. Figure 62a is from Jepson, *Manual of the Flowering Plants of California*, UC Press. Others are originals or redrawings by Bruce Stewart, or are from the original seashore plant books. Photo credits: D. C. Barilotti, pl. 12d; B. Harger, pls. 3c, 3d; L. McMasters, pls. 10c, 10d; F. Menaugh, pl. 4c; M. Neushul, pl. 10b; B. Smith, pl. 4b; J. West, pls. 1c, 1d. All other photos by the authors.

University of California Press
Berkeley and Los Angeles, California
University of California Press, Ltd.
London, England
Copyright © 1982 by The Regents of the University of California
Library of Congress Cataloging in Publication Data

Dawson, Elmer Yale, 1918–1966.
　Seashore plants of California.
　(Natural history guides)
　Combination and revision of Seashore plants of northern California and Seashore plants of southern California / E. Yale Dawson. 1966.
　Bibliography: p. 215.
　Includes index.
　1. Marine flora—California.　2. Coastal flora—California.　I. Foster, Michael S.　II. Title.　III. Series.
QK149.D297　1982　581.9794　81-19690
ISBN 0-520-04138-0　　　　AACR2
ISBN 0-520-04139-9 (pbk.)

Printed in the United States of America

1　2　3　4　5　6　7　8　9

Contents

Preface   vii
1. Introduction   1
    Marine Algae and the California Coastal
      Environment   4
    Plant Distribution in the Intertidal Zone   9
    Plant Distribution in the Subtidal Zone   13
    Collecting and Mounting Seaweeds   16
    Algal Structure and Reproduction   18
    How to Identify Seashore Plants   25
2. Keys to Major Groups of Seaweeds   27
    How to Use a Botanical Key   27
    Color Key to the Major Groups of
      Seaweeds   28
    Key to the Common Genera of
      Green Algae in California   29
    Key to the Common Genera of
      Brown Algae in California   30
    Key to the Common Genera of
      Red Algae in California   36
3. The Green Algae (Chlorophyta)   47
4. The Brown Algae (Phaeophyta)   58
5. The Red Algae (Rhodophyta)   95
6. Sea Grasses   158
7. Coastal Salt Marsh Vegetation   162
8. Coastal Dune Vegetation   174
9. Synopsis   189

Glossary	207
Bibliography	215
Index	219

Preface

This book is a combination and revision of two previous UC Press Natural History Guides by E. Yale Dawson, *Seashore Plants of Northern California* and *Seashore Plants of Southern California*. The considerable overlap in species between the two parts of the state plus the increasing interest in all areas of the coast suggested that a single guide to the seaweeds and common flowering plants found along the shore was most appropriate. The numerous taxonomic changes made since 1966 have been included in the revision and several additional species and much new natural history information have been added.

Drafts of the introduction were read by G. Cailliet, the algae sections by I. A. Abbott, D. C. Barilotti, M. Neushul, and J. Stewart, the sea grass, salt marsh, and dune plant sections by M. Barbour, and the entire text by V. Breda, P. Gabrielson, P. Lebednik, and J. West. I am indebted to all of these friends and colleagues for their many helpful suggestions, to Rosie Stelow for typing the manuscript, and to Lucy Kluckhohn for her excellent editing. Much descriptive and distribution information was adapted from *Marine Algae of California* (Abbott and Hollenberg, 1976).

Dr. E. Yale Dawson died in 1966, the year the original seashore plant books were published. His contributions to marine botany were enormous and an inspiration to all. Dr. Dawson's books communicated information as well as enthusiasm for marine plants, and I hope to have continued at least a small part of his tradition.

Michael S. Foster
June 1979

Introduction

California is not only the most populous but the most popular state of the nation, and its seashore is the playground, the delight, and the inspiration of millions. From La Jolla and Monterey Bay to Moonstone Beach and the rocky islets of Trinidad, our coast is a world attraction of natural beauty. With an increasing awareness of the environment and more time for recreation, the California shore has become familiar to many, and its curious plants and animals have become a source of widespread interest.

Even the most casual visitor to the seashore cannot fail to notice the plants, even though the smells and beach flies associated with decomposing masses of vegetation washed from the sea may create some negative impressions. But many shore animals rely on drift plants for food and shelter, and these natural "compost piles" are recycling centers for the coastal environment. Close inspection of the drift or its sources, the attached vegetation on the rocks at low tide or in the underwater kelp forests, will reveal plants with a diversity of form, structure, and color found in few other habitats on earth.

Unfortunately, although marine plants receive more and more publicity, few books are available to help the layman either to recognize and identify the many interesting and colorful kinds found, or to appreciate their natural history. This book is intended to portray the common and widespread species that occur along the coast in a way that will make most of them easily identified without the need of microscopes or the study of technical descriptions. You will, of course, find plants that are not pictured here, but those presented have been selected from long experience with the marine vegetation of this coast as the ones most frequently encountered. The explanatory notes will help with some of the less common species, and the bibliography provides sources of detailed information for those who wish to take a more advanced step into marine botany.

The majority of the book is devoted to seaweeds, the large, multicellular marine algae or marine macroalgae that are the most conspicuous plants of open coast rocky shores. The term *algae* is commonly used to designate a large, very diverse group of plants from several different kingdoms and phyla. Their common characteristics include structures for sexual reproduction in which all cells are fertile (produce gametes), and an overall anatomy that is relatively simple. Some algae, however, such as the blue-greens, have no sexual reproductive structures; others, like the kelps, have a rather complex anatomy that includes specialized cells, similar to the phloem of so-called higher plants, for conducting sugars. All these plants are now classified using characteristics such as cellular organization and pigmentation. People who study algae are called *algologists* (Latin, *alga*, seaweed; Greek -*logy*, study of), or, preferably, *phycologists* (Greek, *phykos*, seaweed). Although commonly used, the former title is not preferred because it combines roots from two different languages and the Greek root, *algos*, also means pain!

The ubiquitous microalgae that are generally unicellular and microscopic (pl. 1*c*, 1*d*) are not discussed here. These tiny plants occasionally produce macroscopic effects such as green-colored tide pools and the so-called red tides along the shore, but are most abundant in the open ocean where they constitute the bulk of the *phytoplankton* (small wandering plants). Special techniques and equipment are needed for their collection and observation. Some forms of these generally planktonic algae live attached to various substrata, including larger plants, and microscopic examination of the surfaces of larger algae will reveal a diverse "flora on a flora." Other attached forms are colonial and form rather complex, macroscopic structures that commonly resemble filamentous brown algae such as *Giffordia* (p. 58). The two can be distinguished microscopically; the colonial forms are not truly filamentous, but are composed of masses of small cells held together by a gelatinlike substance.

Several generally multicellular but microscopic blue-green algae are also found, both along the open coast and in estuaries. These photosynthetic organisms, more closely related to bacteria than to other types of algae, often form a dark band on high intertidal rocks (pl. 2*b*) and extensive mats on estuarine mud flats (pl. 11*d*). Blue-green algae are common and particularly important because, like bacteria, some have the ability to transform atmospheric nitrogen into chemical forms that can be utilized by other plants. Mixed with microscopic green algae, they frequently give intertidal rocks and shells a greenish discoloration. Some forms can even penetrate rocks, and chipping away a bit of the substratum will reveal a greenish-blue layer of these small plants up to 1 cm (0.4 in) beneath the surface. Even experts have difficulty identifying blue-green algae, and the components of this association can only be determined after laboratory culture.

The generally microscopic terrestrial algae that often give a green coating to tree bark and fence posts near the shore are not treated here. The orange *Trentepohlia*, however, is worthy of mention because of the distinctive coloration it sometimes adds to rocks and wood near the coast, especially the cypress trees growing on the Monterey Peninsula (pl. 1*a*). The green chlorophyll of the plant is obscured by orange pigments, and the densely packed filaments produce a feltlike covering on branches exposed to coastal salt spray and fog. These coverings may be mistaken for lichens, another group of generally terrestrial plants formed from the symbiotic combination of an alga and a fungus. Lichens sometimes form a significant and colorful part of the vegetation on rocks immediately above the high intertidal zone (pl. 1*b*), but require microscopic examination and even chemical tests for proper identification.

Fungi are also important components of marine communities and are actively involved in drift decomposition. There are no truly marine mushrooms, however, and the marine flora is represented by generally microscopic forms. Large fungi are found in the coastal strand and may be identified with the aid of other books in this series.

Like the large fungi, liverworts, mosses, and ferns do not occur in marine habitats, although they can be found near the coast on back dunes, sea cliffs, and in the very high marsh. Flowering plants do occur, both submerged and as the most conspicuous elements of the high beach, dune, and coastal salt marsh flora. The most common species are discussed in this book, but the serious observer of these "higher plants" along the seashore is encouraged to consult the bibliography for more complete treatments.

Marine Algae and the California Coastal Environment

The coast of California is more than 1450 km (900 miles) long and includes coastal habitats ranging from protected, muddy bays and estuaries to rocky headlands that experience the full force of the sea. The California current sweeps down from Washington and Oregon bringing its cold water close to shore as far south as Point Conception (fig. 1). South of Point Conception the California current moves offshore and this, plus the east-west trend in the coast, reduce the influence of cold, southward-flowing water on southern California shores. Strongly influenced by seasonal currents from the south and warmed in slow moving eddies, surface waters in southern California range in temperature from 13° to 20° C (55–68° F), while along the central and northern coast, temperatures of 10° to 13° C (30–55° F) are common and coastal waters rarely get above 15° C (59° F). These currents, plus our wind patterns, can combine to move surface water offshore. The relatively warm surface water is replaced by deeper, colder water, rich in the nutrients essential to plant growth. This process, called upwelling, greatly enhances the productivity of coastal marine algae.

All of the above factors combine to make the marine algal flora of California one of the richest in the world, with more than 280 genera and 660 species currently recognized. These range in form from the primarily tropical *Jania* (p. 109) and *Pachydictyon* (p. 65), most abundant in the relatively warm waters around San Diego, to genera such as *Alaria* (p. 81) and *Constantinea* (p. 103), found in the colder waters north of Point Conception.

These broad geographic influences on the vegetation are superimposed on local factors that can produce quite striking changes along a short stretch of shore. Some of the most obvious changes can be observed while walking from a rocky, wave-exposed point into a quiet bay. Depending on geographic location, plants such as *Postelsia* (p. 77), *Lessoniopsis* (p. 78), and *Phyllospadix* (p. 161) are gradually replaced by genera such as *Iridaea* (p. 133) and *Gracilaria* (p. 121). Although correlated with changes in water motion, however, one must be careful in assuming that differences in ability to cling to the substratum are the cause of these changes. Quiet-water species are generally more delicate and likely to be torn loose on exposed points, but there are no obvious physical or chemical reasons why a plant like *Postelsia* should not grow in calm water. Changes

Fig. 1. Map of California showing coastal landmarks and cities referred to in the text

in grazing animal species and grazer abundance, as well as competition for space and light, also occur along exposure gradients, and unraveling the relative importance of all of these potential factors on algal distribution is a fruitful area of research in marine ecology.

Even more striking differences in the algal flora can be observed while traveling from the open coast into an estuary like San Francisco Bay. In addition to all the changes discussed above related to reduced water motion, salinities also drop as salt water mixes with the fresh water from rivers and streams flowing towards the sea. Few seaweeds can make the necessary adjustments in their water and mineral content to survive at low salinities, and the number of species rapidly declines inland until only plants such as *Gracilaria* (p. 121) and *Enteromorpha* (p. 51) are left. Even these disappear at very low salinities. Many small estuaries can, at certain times during the year, become almost completely filled with fresh water if their openings are blocked by sand movement. In contrast, if tidal circulation is reduced and streams stop flowing as they frequently do during California summers, evaporation of the salt water can lead to hypersaline conditions that are also intolerable to most large algae.

Local differences in substratum are also important, and hard, difficult-to-erode rocks such as granite support some of our most diverse local floras. Perhaps the greatest local differences are found in sandy versus rocky areas. Considerable amounts of drift algae from nearby rocks may occur on sandy beaches but, lacking a place to attach, no seaweeds grow in areas permanently covered by sand. This is one of the reasons for the relatively poor algal flora on the Atlantic coast of the United States, dominated as it is by vast, uninterrupted stretches of unstable and abrasive sand. Many sandy beaches in California are seasonal, however; sand is moved offshore by large winter waves and moved back onto the beach by the small waves of late spring and summer. Thus, solid substratum is available if an alga can withstand abrasion and complete burial for long periods, or if it can establish, grow, and reproduce during the brief periods when rock is exposed. The flora of such areas is greatly reduced, but a surprising number of algae are adapted to these conditions and some are found almost nowhere else.

In addition to these floral changes related to natural differences in physical and biological factors, man continues to have a dramatic effect on coastal vegetation. The creation of breakwaters, harbors, and other structures along the coast, combined with dredge and fill activities, and a reduction in river sediment transport below dams have caused extensive alterations in intertidal and subtidal substrata. In general, marshes have been filled and beach and coastal bluff erosion has increased. In some areas, habitats that were once rocky are now sandy beaches. Such changes have completely altered, and even destroyed, local plant communities.

Waste discharges, especially along the heavily developed coast from Los Angeles to the Mexican border, have been implicated in an overall reduction of the intertidal flora and in the loss of offshore kelp forests. Work in both California and Japan has also shown that air pollution may affect the composition of the intertidal flora. Although the algae appear relatively resistant to crude oil pollution, surf grasses and marsh plants are not, and we can expect further damage to these plants if oil spills continue.

Unfortunately, the increase in human numbers, the increased interest in coastal natural history, and greater and greater use of the coast for recreation have also created severe problems for open coast intertidal zone, marsh, and dune organisms. Thousands of feet walking over the same area can destroy local communities and, in dunes, create tremendous erosion problems. Additional disturbance caused by removing plants and animals (which is against the law without a permit; see p. 16) and turning over rocks only increases the problem. Although the coast is long, the organisms along it constitute a very thin line that could be easily erased by overuse. We must make every effort to ensure that our use and enjoyment of coastal communities does not inadvertently cause their destruction.

Plant Distribution in the Intertidal Zone

In addition to those mentioned above, factors related to tidal action have a profound influence on the distribution of intertidal marine plants. Two high and two low tides occur along the coast during approximately every 25-hour period. The magnitude of the tides varies during the year, with the highest and lowest occurring around January and June. During these periods, the water level may change as much as nine feet (2.8 m) in about six hours, and this nine-foot vertical range defines the intertidal zone along the California coast. The average of the lowest low tides is defined as Mean Lower Low Water. A tide occurring below this point is called a minus tide (listed in tide tables in feet with a minus sign in front), and tides above this point are called plus tides. During the extreme tidal periods, minus tides may occur to minus two feet (0.6 m) and plus tides to seven feet (2.1 m), giving the nine-foot total change.

Being covered and uncovered by water can have a profound influence on the algae; while out of water, they may experience lack of essential nutrients, desiccation, increased radiation, and increased or decreased temperatures and salinities. For marine organisms, the intertidal zone thus represents a physical-chemical stress gradient with almost completely marine, low-stress conditions in the low intertidal zone and almost completely terrestrial, high-stress conditions in the high intertidal zone.

Analogous gradients of temperature and rainfall exist between the floor of the central valley and the top of the Sierra. The broad zonation pattern of plants, with

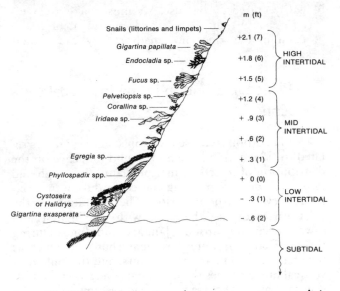

Fig. 2. Vertical zonation of some common seaweeds in the rocky intertidal zone

grassland grading into oak woodland and then into various conifer communities, reflects the direct and indirect effects of these gradients on terrestrial vegetation. This zonation in response to climatic change also occurs on the shore, but is compressed into nine vertical feet, rather than the thousands of feet up the Sierra. Certainly the most striking aspect of rocky intertidal areas at low tide is the rather distinct vertical zonation of various species (pls. 2b, 2c; 3a). A greatly simplified, composite pattern for the entire state is shown in figure 2. In any particular locale, species generally occur within a rather small vertical range along the gradient. Although species change in abundance and drop out, new species are added as one moves up or down the coast and vertical zonation still occurs.

Since the plants are not scattered about at random, it is important to consult a tide table (available at most sporting goods and fishing supply stores) before going to the shore to observe seaweeds. Unless it is an extremely high tide, some plants will always be exposed, but lower levels tend to be richest in species, and during low tides more of the entire gradient is exposed for observation.

Changes in tidal exposure can affect the distribution of intertidal organisms both directly and indirectly. Research with several intertidal animals and a few seaweeds has shown that the upper distributional limits of most do result from their inability to tolerate further exposure to terrestrial conditions. Species differ in their ability to tolerate exposure and thus live at different places along the gradient. But experiments indicate that the lower limits are often not directly related to some physical or chemical change associated with the tides, and many intertidal organisms actually grow and reproduce better if transplanted below their normal range. It appears that lower limits are generally the result of a particular alga's inability to compete for space or light with species growing below it, or inability to withstand grazing from animals living below. Thus, rather than representing an optimal environment, intertidal plants may actually be forced to live in refuges.

Imagine an alga that is grazed by a limpet that is preyed upon by a sea star whose upper limit is determined by tidal exposure and you will begin to see how complex the relationships among tides, algal growth, and interactions with other organisms can be. Although the relationships are complex, a careful observer can discover many of these interactions, and a much greater appreciation of the intertidal zone as a community can be gained by observing the animals associated with the seaweeds. A further discussion of tides and intertidal animals can be found in the other books in this series, for example, by Hedgpeth and Hinton (see the bibliography).

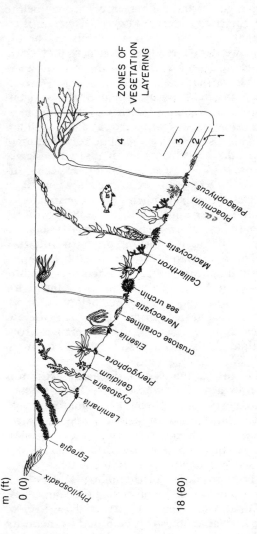

Fig. 3. The distribution of some common subtidal seaweeds. Four vertically arranged layers of vegetation include:
1. Turf species such as encrusting corallines and *Pterosiphonia* spp.; 2. Plants up to 0.5 m (20 in) tall (*Gelidium, Calliarthron, Plocamium*); 3. Plants up to 2 m (6.5 ft) tall (*Eisenia, Pterygophora, Laminaria*); 4. Midwater and surface canopy species such as *Egregia, Cystoseira,* and *Macrocystis*

Plant Distribution in the Subtidal Zone

The bulk of the seaweeds along the coast occur in the subtidal zone, that region below the lowest reach of the tides. As in the intertidal zone, plants growing here are also influenced by variations in temperature and nutrients, as well as local differences in substratum and water motion. They do not face all the variations in climate found in the periodically exposed intertidal zone, but gradients of other factors can have a profound effect on their distribution. Although not nearly as distinct as in the intertidal zone, there is a vertical zonation of species with depth. Figure 3 shows a composite distribution of several more common subtidal species in California.

Plants living in the subtidal zone encounter changes in light not found in intertidal areas. As depth increases, light intensity drops rapidly and light quality shifts from the spectrum of full sunlight to blues and greens. These changes are known to limit the vertical distribution of several macroalgae, and few seaweeds can grow at depths greater than 50 m (164 ft) in the moderately turbid waters of California. The subtidal habitat is usually more three-dimensional than the intertidal habitat, and plants with floats or stiff stipes can extend up in the water and further alter the light regime of plants growing beneath them. This situation is similar to that found in terrestrial forests, where the overstory trees alter the light reaching smaller trees, shrubs, and other understory plants. Subtidal vegetation may occur in several layers (fig. 3), producing quite complex changes in light in addition to those caused by the water alone. The effects are especially obvious under some giant kelp forests where the shade produced by a very dense surface canopy can greatly reduce the growth of understory algae.

Water motion, like light, also decreases with depth. Although the most obvious power of waves can be seen in the crashing surf, tremendous forces are also generated prior to breaking as waves moving into shallow water "feel the bottom" before reaching the shore. The back-and-forth surge produced increases in speed as water depth decreases. In addition to tearing plants from the substratum, surge can make the observation of subtidal plants difficult. Strong surge can pull the mask from a diver's face, and one quickly learns which plants cling most tenaciously to the bottom, for these provide convenient handholds to prevent being washed over the bottom. Plants such as the southern sea palm (p. 83), feather boa kelp (p. 75) and surf grass (p. 161) can withstand these forces, while giant kelp is not as well adapted. The former species are generally most abundant from the low intertidal zone down to about 6 m (20 ft), while giant kelp does best in deeper water (fig. 3).

Most intertidal and subtidal habitats are characterized by a very high cover of plants and animals, suggesting that space is also an important limiting resource. Seaweeds must not only compete with each other for space, but with attached animals such as sponges and bryozoans. New space may become available through a variety of processes including the removal of organisms by waves and surge, and the removal of sand by various types of water motion. This space is quickly colonized by planktonic spores and larvae, and competition for space often plays an important role in determining the species composition of the new association.

Large grazing animals such as sea urchins can also have an impact on marine vegetation and are thought to be responsible for destroying some giant kelp forests in southern California. These animals, however, generally feed by "catching" drift algae and usually do not actively move and forage on attached plants unless the drift supply is low. If natural predators such as sea otters, certain fish, and even lobsters and crabs are reduced, sea urchin populations can increase, reduce the available drift, and then begin to overgraze their surroundings. Drift algae can also be directly reduced if plant growth is inhibited by physical and chemical changes in the environment. These environmental changes, which can result in overgrazing, may occur naturally, but man has also created them by overharvesting sea urchin predators and by polluting coastal waters.

Giant kelp forests are among the most, if not the most, productive plant communities in the sea. Most of this plant production is not consumed immediately, but enters local food webs as drift. Moreover, less than half of this drift is actually consumed in the forest; the rest is either moved onto the beach where it supports intertidal organisms, or it is moved offshore. Drift transported seaward along the bottom moves into deep water and even down into the depths of submarine canyons. Given the amounts of material moved, drifting plants from kelp forests may be a very important source of food for animals living in deep water where lack of light prevents plant growth.

Plants with hollow parts may float out of the forest as surface drift, referred to as kelp rafts or patties. Moved by winds and surface currents, these floating islands can be found far out to sea where they provide habitats for a variety of juvenile and adult fishes. If conditions are right, the plants survive and may even continue to grow in the surface drift, and such rafts may be important seaweed dispersal agents.

In addition to providing food and habitats for numerous mobile animals, intertidal and subtidal algae provide substrata for a vast number of sessile animals. The time spent carefully observing the surface of almost any large seaweed will be well rewarded by the discovery of the myriad animals such as worms, hydroids, and bryozoans living on them and in their holdfasts.

Collecting and Mounting Seaweeds

Wisely, California has established laws regulating the collection of marine life. Attached seaweeds to be used for food may be collected, but the collector must have a sport fishing license. Collection of attached plants for other purposes is forbidden out to 914 m (1,000 yds) offshore unless one has a special collecting permit. The serious seaweed collector can apply for such a permit through the California Department of Fish and Game in Sacramento. Even with a permit, all collecting is forbidden in state parks and ecological reserves without special permission from the managing agency. Drift specimens, however, may be collected without a fishing license or collecting permit. Even without laws, we feel that the collection of live, attached marine and terrestrial plants should be avoided for other than scientific and educational purposes. Although no marine algae are presently endangered or even threatened, their unnecessary removal degrades the environment for other plants and animals (including man) and also results in the removal of additional plants and animals that may be living on the algae. "Collecting" algae on film with a camera can provide an excellent record of your observations and includes a visual record of other aspects of the habitat without damaging it.

Seaweeds make beautiful pressings when properly mounted. Fortunately, most can be found in good condition in beach drift where their removal creates little disturbance. If you are going to collect drift plants for later identification, try to find entire specimens, including holdfasts. If collected plants cannot be mounted immediately, they can be preserved in a suitable container. It is easiest to collect specimens in plastic bags and to preserve them in a mixture of seawater and formaldehyde (about 19 parts seawater to 1 part formaldehyde). Formaldehyde is toxic; handle it with care in a well-ventilated room or outdoors. If the specimens are to be kept for some time before mounting, they should be placed in a tight metal can or, if in glass, at least in a very dark place to prevent bleaching.

Mounting is also easy, but requires a few special materials. Plants are prepared by spreading and pressing them onto a piece of high-quality rag paper. Most will eventually stick by their own mucus, but some must be glued after drying. Regular herbarium paper may be obtained from Carpenter/Offutt Paper, Inc., P.O. Box 3333, South San Francisco, CA 94080. A standard plant press, cardboard ventilators, and drying felts should be at hand.

The fresh or preserved seaweeds are floated out in water in a broad pan or sink and the paper to be used as backing placed under them in the water. If preserved, tap water may be used. Use only seawater with fresh specimens, as tap water will cause the pigments in some algae to leach. Only a very small amount of water is needed to wet the paper and to spread and arrange the specimen. The paper is then carefully withdrawn with its specimen, drained momentarily and laid on a drying felt on top of a ventilator in the press. It may then be covered with a sheet of household waxed paper or a piece of cotton sheet and another drying felt and cardboard placed on top for the next specimen. When the pile of materials is complete, the press is strapped up and the drying allowed to proceed. The wet felts should be replaced with dry ones at least once a day until the specimens are dry. The back of a refrigerator or any warm spot indoors or out will hasten the drying process. Long drying times encourage the growth of mold on fresh specimens.

After the specimens are dried on their backing sheets, they may be provided with appropriate labels indicating place, date, and conditions of collecting. If professional supplies are not available, almost any thick, white paper will work; the ventilators can be eliminated and newspaper substituted for felts. Be sure to take care that the newsprint does not contact the white paper. Two boards and heavy straps or weights can be used in place of a standard press. Above all, take care in arranging your specimens on the paper before drying. If you are patient, the results will be aesthetically, as well as botanically, pleasing. Contact the biology department of your local college or university if you have any problems or need to locate supplies.

Many marine algae are delicate and have characteristics that can only be determined microscopically. For purposes of this book, however, attention will be confined to the larger forms, which can usually be recognized by their gross characteristics.

Algal Structure and Reproduction

Most of the macroscopic seaweeds belong to three principal groups (phyla) and these are known by the predominant colors assumed by many of their members. Thus, we have the green algae (Chlorophyta), of which most members are of a distinctly green color (p. 47, pls. 1e, 2a, 6a), the brown algae (Phaeophyta), which are generally brownish (p. 58, pls. 3c, 3e, 4c, 5a) and include the large kelps, and the red algae (Rhodophyta), which are often reddish, but may have a mix of pigments that lend a purple, pink, yellow, or brown color (p. 95, pls. 7a, 8a, 8c, 10e).

Almost all algae are spore-producing plants, and the spores are microscopic. Most reproductive organs, in fact, are visible only with a microscope, but for critical identification of species, they frequently must be examined. Some reproductive structures are sufficiently conspicuous that they are mentioned in the following pages, and examination of these and other plant parts with a simple hand lens will reveal many interesting and distinctive features invisible to the naked eye.

Algae generally have complex life histories that

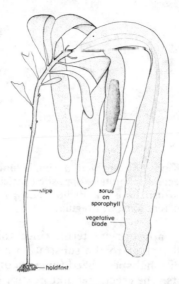

Fig. 4. *Pterygophora californica*, a young plant 50 cm (20 in) tall

include an alternation of a sexual, gamete-producing haploid generation with a spore-producing diploid generation. Thus a single species may consist of three individual plants, a male gametophyte, a female gametophyte, and a sporophyte (fig. 8, below). These individuals commonly resemble one another, but in some species, such as the kelps, the sexual plants are minute and observable only microscopically (fig. 9a, pl. 4b). In a few species, both the sexual and spore-producing plants are macroscopic but dissimilar and look like different species. In fact, as a result of culture studies, many plants once classified as separate species have in recent years been recognized as simply different phases in a single life history.

Algae have no true roots, leaves, or flowers (fig. 4). Although parts of many larger seaweeds resemble roots or leaves, we treat the entire algal body as a *thallus*. The attaching portion is called the *holdfast*,

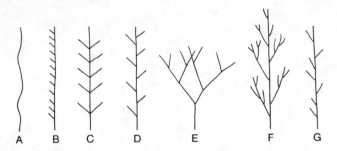

Fig. 5. Kinds of branching: *a*, simple (unbranched); *b*, pectinate or secund; *c*, pinnate opposite (distichous); *d*, pinnate alternate (distichous); *e*, dichotomous; *f*, monopodial showing percurrent axis; *g*, irregular

an especially appropriate term, since this structure only holds the plant to the substratum and does not have the additional specialized function of absorption found in roots. The erect, stemlike portion is called the *stipe* and the leafy parts are the *blades*. In some algae spores are produced only on specialized blades called *sporophylls*. The spore-producing structures are frequently aggregated into *sori*, which generally appear as dark areas on blades.

Some algae have finely dissected or bladelike thalli in which neither stipe nor blade can be recognized. Almost all have holdfasts, however, unless the plant has broken loose. Except for some green algae with multinucleate, noncellular structure, the seaweeds are composed of definite cells, which often have a precise and distinctive shape and arrangement. Many of the identifying characters of the algae are to be found in the cell structure. Some simple forms consist of a single branched row of cells; some, of a sheet of cells in one or two layers. Some are composed mainly of similar-sized cells while others have tissues made up of different types and sizes of cells. The cell walls may be thin and definite, or sometimes very thick, gelatinous, and indefinite. In seaweeds that are many cells

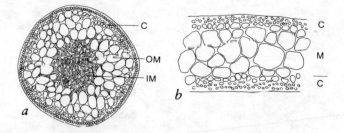

Fig. 6. Internal cellular arrangement in two red algae. *a*, *Neoagardhiella gaudichaudii* (3 mm, 0.12 in); *b*, *Rhody-menia* sp. (0.5 mm, 0.02 in); *C*, cortex; *IM*, inner medulla; *OM*, outer medulla; *M*, medulla

thick, the thallus in cross section is made up of an outer layer of pigmented cells called the *cortex*, and an inner area of generally nonpigmented cells of different size and/or shape called the *medulla* (fig. 6). The arrangement of branches on the thallus may also be distinctive and is extensively used as a key characteristic (fig. 5).

Spores produced by algae are of two principal kinds, motile and nonmotile. In the green and brown algae, the spores are usually motile, flagellated unicells called *zoospores* (figs. 8, 9*a*). They are produced in various ways and positions in *sporangia*. In the red algae, spores are nonflagellated and one type is usually produced in groups of four *tetraspores*. These occur in several distinct arrangements useful in classification (fig. 7).

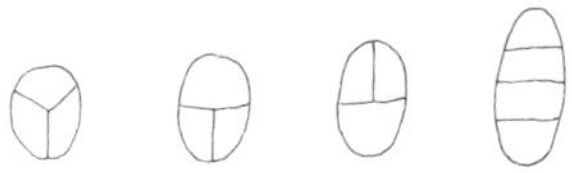

Fig. 7. Some arrangements of tetraspores in tetrasporangia: left to right, tetrahedral, cruciate, cruciate, zonate. Where only three division planes are shown, the fourth is out of view.

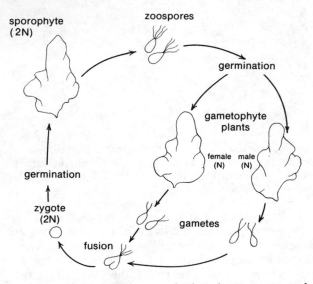

Fig. 8. The life history of *Ulva*, showing gametophyte and sporophyte stages

As indicated above, the spores produced by sporophytes usually give rise not to more sporophytes but to sexual plants (gametophytes), often male and female. In green (fig. 8) and brown algae (fig. 9a), the sexual plants for the most part produce motile, flagellated gametes, often similar to the zoospores. Male and female gametes may be similar or different in size, or the female may be a nonmotile egg and be fertilized by a much smaller male gamete. Some algae, like *Fucus* and other rock weeds (fig. 9b), do not produce spores, having an animallike life history. Gametes of the red algae are all nonflagellated. The small male *spermatium* fertilizes a female gamete that does not leave the female plant, and the zygote undergoes a complex development in the female thallus before producing another

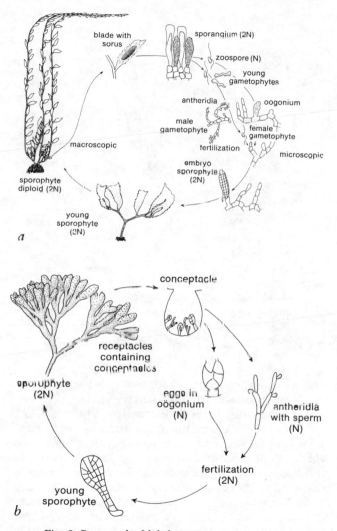

Fig. 9. Brown algal life histories: *a*, *Macrocystis* (giant kelp); *b*, *Fucus*

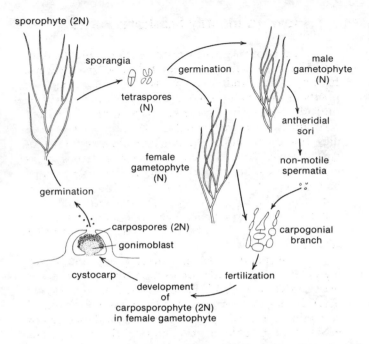

Fig. 10. The life history of the red alga, *Gracilaria*

kind of spore, called a *carpospore*. This spore is released and produces the sporophyte (fig. 10).

Examination of cellular structure and other fine details in algae requires the preparation of microscope slides bearing small portions or cut sections of plants which can be examined with a compound microscope. This kind of investigation may be carried out by those who wish to go beyond the scope of this introductory book. More extensive collection and study may be undertaken through the use of technical literature listed in the bibliography. Most colleges and universities have courses dealing with the biology of algae and other seashore plants, which can be taken by anyone.

How To Identify Seashore Plants

The best way to identify seashore plants is to take this book with you in the field and use the keys, figures, plates, and descriptions while observing the plants in their natural habitat. Pieces of plant may be difficult to identify if some key part is missing, so observe an entire plant (including the holdfast if it is an alga) and try to find one that is reproductive. In the algae, reproductive structures generally appear as bumps, dark spots, or patches on the thallus. Size must be determined to identify some species, so take a ruler with you. The longest dimension of each plant illustrated is given in the figure legends. In addition, a 10x hand lens will aid in observing some of the smaller structures.

If the plant is lax, grasslike, green, and growing in the low intertidal or subtidal zone, turn to the sea grass section (p. 157) and examine the figures and descriptions to determine which plant you have. If you want to identify plants from coastal dunes or salt marshes, look at the figures and read the descriptions in the appropriate section (dune plants, p. 174; marsh plants, p. 162). Try to find a flowering specimen.

Only very common dune and marsh plants are included in this book, so keys are not provided for them. For the algae, however, use the keys provided to avoid turning a lot of pages. By carefully observing the characteristics of your seaweed and following the keys, you should quickly arrive at one or a few possible genera. A final decision can be made after reading the appropriate descriptions and examining the figures or plates. The distribution of each alga in California is given at the end of each description. Unfortunately, few algae have widely accepted English common names. In countries where seaweeds are used extensively, almost all plants have common names, but this is not the case in the United States. Therefore, you will generally end up with only a scientific name for your plant. Latin and Greek translations are given to help make sense out of these often long and difficult-to-pronounce names.

2

Keys to Major Groups of Seaweeds

How to Use a Botanical Key

A botanical key is a guide to finding the name of a plant by making a series of choices between statements about it. In this book, the statements are arranged in pairs and you select from the two statements in the pair the one that best describes the alga you wish to identify. Listed to the right of each statement is the name (genus) of an alga or the number of the next pair of statements to consult. The process of choosing between statements is repeated until you come to a name or arrive at the last plant in the key. After reading the text and looking at the illustrations of plants in the genus, you may find that your plant does not fit the description given. If so, you may have made a wrong choice and should try the key again. If repeating the key does not help, you probably have a plant not treated in this book or an atypical specimen and you should consult one of the books listed in the bibliography or examine the illustrations for something similar. Be sure to read the additional notes given with some of the statements, as they have been added to eliminate the most common sources of confusion.

As an example of how to "key out" a plant, let's assume you want to identify some small red blades (*Smithora*, p. 98) attached to surf grass (p. 161). Examine the illustration (fig. 27c), read the description (p. 98), and then follow along through the keys as if you didn't already know the name. First, you would turn to the Color Key to the Major Groups of Seaweeds on p. 28. Starting at pair 1, the plant is red, not green, so the second statement is correct and the right-hand column tells you to go on to pair 2. The second statement of pair 2 is correct for your plant and the right-hand column tells you it is a red alga and to proceed to the Key to the Common Genera of Red Algae on p. 36. Again, starting at pair 1, your plant is "not stony," the second choice, so go to pair 9. The rest of the sequence, based on continued comparisons of the plant with the descriptive statements, is 9-11-15-18-27-28-30-31. The first statement in pair 31 is correct and the right-hand column tells you that the plant is *Smithora* and to turn to p. 98 for a discussion of this genus.

Color Key to the Major Groups of Seaweeds

1. Color distinctly green (grass green, dark green, or yellow green) .... Green Algae (Chlorophyta)
   Key—p. 29
   Descriptions—p. 47

   NOTE: *The coarse, branched rockweeds (*Fucus, Pelvetia, Pelvetiopsis, Hesperophycus*) are brown algae, but frequently appear olive-green (pl. 5b). If your plant is like this, go to the brown algae key, p. 30. The branched red alga,* Gastroclonium *can also be green, but is otherwise distinguished by its small, hollow branchlets (p. 135, pl. 9d). Blades of the inter-tidal red alga,* Iridaea flaccida *(p. 133) are also generally green in appearance and can be confused with the green alga,* Ulva *(p. 52); the* *blades of Iridaea, however, are thicker and more rubbery than Ulva, have red spores, and have an oily iridescence when submerged. The hollow thallus of the red alga, Halosaccion (p. 136), may also be green, but generally has some reddish color near the base.*

1. Color not distinctly green 2

2. Color distinctly brown (dark brown to yellowish brown or almost black, but not reddish) .... Brown Algae (Phaeophyta)
    Key—p. 30
    Descriptions—p. 58

    NOTE: Some red algae, particularly those in the high intertidal zone like Gigartina papillata (p. 129), Rhodoglossum affine (p. 131), and Endocladia (p. 114), may appear brownish or black. But they usually have a reddish tinge or some red areas and always have red spores.

2. Color generally red or reddish, but variable, sometimes pink, or very dark, dull red to almost black; rarely green ....
    .... Red Algae (Rhodophyta)
    Key—p. 36
    Descriptions—p. 95

Key to the Common Genera of Green Algae in California

1. Plants hollow, tubular ... *Enteromorpha* (p. 51)
    Plants otherwise 2

2. Plants spherical .... *"Halicystis"* (p. 56)
    Plants otherwise 3

3. Plants thick and spongy, dichotomously branched or cushion-shaped .... *Codium* (p. 54)

   Plants foliose or filamentous 4

4. Plants foliose .... *Ulva* (p. 52)

   Plants filamentous 5

5. Plants featherlike; filaments without cross walls (siphonous) .... *Bryopsis* (p. 51)

   Plants not featherlike; filaments with cross walls 6

6. Filaments branched 7

   Filaments unbranched 8

7. Filaments with hooked or spinelike branches, entangled in strands .... *Spongomorpha* (p. 50)

   Filaments without hooked branches; frequently forming mosslike cushions ....

   .... *Cladophora* (p. 49)

8. Cells in filament large (usually greater than 0.5 mm [0.02 in] long) and generally visible to the naked eye .... *Chaetomorpha* (p. 49)

   Cells small and not visible to the naked eye .... *Ulothrix* (p. 47)

Key to the Common Genera of Brown Algae in California

1. Plants filamentous, tufted, and usually less than 5 cm (2 in) tall (colonial diatoms often have a similar form; see p. 3) .... *Giffordia* (p. 58)

   Plants larger and firmer in structure; not composed of free filaments 2

2. Entire plant crustose 3

   Plants erect (holdfast may be crustose) . . . . , . . 4

3. Crusts firm, generally smooth . . . *Ralfsia* (p. 60)

   Crusts soft, usually spongy and convoluted . . . . .
   . . . . *Cylindrocarpus* (p. 60)

4. Upright portions mostly hollow throughout . . . .
   . . . . , 5

   Upright portions not hollow throughout, although
   sometimes with air bladders or hollow stipes
   8

5. Plants elongate, tubular; tubes often constricted at
   intervals . . . . *Scytosiphon* (p. 63)

   Plants bubble-shaped or saccate 6

6. Plant surface covered with dark brown spots; plant
   epiphytic on *Rhodomela* (p. 155) or *Odonthalia*
   (p. 157) . . . . *Soranthera* (p. 62)

   Plants without dark spots, surface smooth, convoluted, or warty 7

7. Plants saccate, cylindrical, or compressed, epiphytic on *Cystoseira* (p. 90) or *Halidrys* (p. 92) . . .
   . . . . , ... *Coilodesme* (p. 61)

   Plants saccate, globular, frequently convoluted, or
   with warty outgrowths . . . *Colpomenia* (p. 62)

8. Mature plants with one or more hollow air bladders or hollow stipes 9

   Mature plants without hollow parts 17

9. Plants with long, slender, hollow stipes terminating
   in a single, large bulb 10

   Plants with many small air bladders or a hollow
   stipe without bulb 11

10. Blades borne on two large, antlerlike branches attached to the bulb ..... *Pelagophycus* (p. 72)

Blades borne on small branchlets attached to bulb
.... *Nereocystis* (p. 70)

11. Plants with a tough, flexible, hollow stipe terminating in a tuft of drooping blades ....
.... *Postelsia* (p. 77)

Plants with many small air bladders 12

12. Plants dichotomously branched throughout; air bladders at ends of some or all branches (branch tips of *Hesperophycus* [p. 90] may also be inflated) .... *Fucus* (p. 86)

Plants not dichotomously branched, or only so near holdfast 13

13. Plants with a single, large (1–3 cm [0.4–1 in] long) air bladder at the base of lateral blades ... 14

Plants with small (less than 1 cm [0.4 in] long) air bladder in series or solitary on short stalks ...
.... 15

14. Main axis cylindrical ...... *Macrocystis* (p. 73)

Main axis flattened .... *Egregia* (p. 75)

15. Air bladders solitary and borne on short stalks
.... *Sargassum* (p. 92)

Air bladders in series 16

16. Serial air bladders spherical or slightly flattened, beadlike .... *Cystoseira* (p. 90)

Serial air bladders distinctly flattened ....
.... *Halidrys* (p. 92)

17. Plants unbranched (although blades may be tattered; a plant with unbranched blades attached

to an unbranched stipe [like *Pterygophora*, p. 81] is considered unbranched) 18

Plants branched 23

18. Plants without a distinct stipe 19

Plants with a distinct stipe 20

19. Plants usually less than 30 cm (12 in) long; blades thin, flat, and growing in lax tufts ....
.... *Petalonia* (p. 65)
(You may also have a *Coilodesme* [p. 61] but failed to notice it was hollow, or a young, un-branched *Taonia* [p. 68] or *Zonaria* [p. 66].)

Plants usually larger, growing in stiff clumps; blades thick and often convoluted ....
.... *Hedophyllum* (p. 80)

20. Plants with a single, terminal blade 21

Plants with terminal and lateral blades 22
(If your plant has a tuft of terminal blades, you have *Postelsia* [p. 77] but failed to notice the hollow stipe; if it has radially attached blades, it is probably a vegetative *Cystoseira* [p. 90].)

21. Terminal blade with longitudinal ribs ....
.... *Costaria* (p. 85)
(You may have *Dictyoneurum* or *Dictyoneuropsis* [p. 80] but didn't notice stipe branching.)

Terminal blade without longitudinal ribs ....
.... *Laminaria* (p. 84)
(One variety of *Desmarestia* [p. 69] and young specimens of almost all kelps look like this; only *Laminaria* retains the form when mature.)

22. Lateral blades smaller than terminal blade; terminal blade with distinct midrib ... *Alaria* (p. 81)

Lateral blades usually the same size as terminal blade; terminal blade lacking distinct midrib (although axis slightly thickened) . . . . *Pterygophora* (p. 81) (If plant has small lateral blades and no midrib on terminal blade, it is probably a young *Egregia* or old *Egregia* without air bladders [p. 75].)

23. Blades branched; stipe inconspicuous or lacking 24

    Blades unbranched; stipe branched, conspicuous, stalklike, stumplike, or prostrate; plants generally taller than 15 cm (6 in) 33

24. Branching opposite, blades with fine axial veins . . . . *Desmarestia* (p. 69)

    Branching not opposite 25

25. Blades with midribs (especially prominent in older portions) 26

    Blades lacking midribs 28

26. Plants olive to dark brown, regularly dichotomous, with generally swollen branch tips 27

    Plants yellowish-brown to olive, irregularly dichotomous, without swollen branch tips . . . . *Dictyopteris* (p. 68)

27. Blades with tiny (less than 1 mm [0.04 in] in diameter) tufts of white hairs lying in two parallel rows, one on either side of midrib . . . . *Hesperophycus* (p. 90)

    Tufts, if present, not in parallel rows along midrib . . . . *Fucus* (p. 86)

28. Branches markedly flattened (width at least 10 times thickness) 29

Branches not markedly flattened 31

29. Plants fan shaped, apices usually light colored
.... *Zonaria* (p. 66)

Blades more or less strap shaped 30

30. Branch tips rounded ...... *Pachydictyon* (p. 65)

Branch tips plane or tattered .... *Taonia* (p. 68)

31. Plants tan to dark brown; branching polystichous
(in various planes) .... *Analipus* (p. 63)

Plants olive brown, dichotomously branched ....
.... 32

32. Plants forming dense clusters 4–8 cm (1.5–3 in)
tall .... *Pelvetiopsis* (p. 88)

Plants loose and drooping, 20–90 cm (8–35 in)
long .... *Pelvetia* (p. 88)

33. Conspicuous stalklike stipe forked at apex ......
.... *Eisenia* (p. 83)

Stipe not stalklike, dichotomously branched ....
.... 34

(If you got this far and your plant does not fit
33, you may have a *Laminaria* [p. 84] with a
branched rhizome.)

34. Stipes forming a stumplike, woody base; upper
portions dichotomously branched with linear,
smooth, terminal blades ....
.... *Lessoniopsis* (p. 78)

Stipes flattened, mostly prostrate, dichotomously
branched; terminal blades with rectangular con-
volutions .... *Dictyoneurum* (p. 80)

Key to the Common Genera of Red Algae in California

1. Plants stony, calcified (corallines) 2

   Plants not stony 9

2. Entire plant a crust 3

   Part of, or whole, plant erect and jointed (articulated) 5

3. Plants epiphytic 4

   Plants forming crustose masses on rocks or other hard substrata ....
   ..... *Pseudolithophyllum* and others (p. 106)

4. Forming crusts on noncalcareous algae and sea grasses (p. 157) .... *Melobesia* (p. 108)

   Forming disks on articulated corallines ....
   .... *Mesophyllum* (p. 109)

5. All segments (intergenicula) cylindrical 6

   Some or all intergenicula cone shaped or flattened
   7

6. Intergenicula short and about as long as broad (around 0.5 cm [0.2 in]) ....
   .... *Lithothrix* (p. 109)

   Intergenicula longer and two or more times longer than broad .... *Jania* (p. 109)

7. Intergenicula shorter than greatest width, markedly flattened, conceptacles (reproductive bumps) on flattened faces of intergenicula ....
   .... *Bossiella* (p. 112)

   Intergenicula longer than broad 8

8. Conceptacles at tips of unbranched intergenicula
.... *Corallina* (p. 111)

   Conceptacles mostly on margins of intergenicula,
   occasionally on faces in lower parts of plant
   .... *Calliarthron* (p. 113)

9. Plants forming expanded crusts or wartlike cushions 10

   Plants with free parts (may have crustose holdfasts)
   11

10. Forming an expanded crust ....
    .... *Peyssonnellia* and others (p. 105)

    Forming a wartlike cushion on *Laurencia* (p. 154)
    or *Chondria* (p. 151) ... *Janczewskia* (p. 152)

11. Plants filamentous (cells arranged in a column one
    cell broad); may be more than one cell broad in
    upper portions 12

    Plants not filamentous 15

12. Filaments unbranched, upper portions generally
    more than one cell broad ..... *Bangia* (p. 97)

    Filaments branched 13

13. Branching distichous (in two rows along main axis)
    .... *Antithamnion* (p. 137)

    Branching otherwise 14

14. Plants less than 2 cm (0.8 in) tall, forming feltlike
    masses in shaded areas ....
    .... *Rhodochorton* (p. 97)

    Plants generally taller than 2 cm; hairlike ......
    .... *Tiffaniella* (p. 147)
    (If you got this far and your plant is not *Tiffa-
    niella*, it is probably not entirely filamentous.

Examine it carefully and start the key again at 15.)

15. Cells of thallus in regular transverse series (generally appear filamentous without a hand lens; see *Polysiphonia*, fig. 51*a*) 16

    Cells not in a regular transverse series 18

16. Branching not distichous (in the two rows along main axis); usually radial ....
    .... *Polysiphonia* (p. 149)

    Branching distichous 17

17. Axes of main branches percurrent (extending throughout length of plant) ....
    .... *Pterosiphonia*, in part (p. 151)

    Axes of main branches not percurrent ....
    .... *Pterochondria* (p. 151)

18. Plants bladelike; blades with veins and/or midrib (veins may be small; examine with hand lens)
    19

    Plants bladelike or otherwise; lacking veins and/or midrib 27

19. Midrib continuous or nearly so except for uppermost portions of the blade(s) 20

    Midrib not continuous (interrupted, only in basal portion, or absent) 22

20. Midrib continuous except for uppermost portion of blades; blade margins with teeth ....
    .... *Nienburgia* (p. 143)

    Midrib continuous from base to apex; no teeth 21

21. Blades usually branched or with proliferations;

veins often in opposite pairs along midrib
.... *Phycodrys* (p. 140)

Blades unbranched (although frequently with numerous lateral splits), featherlike in shape, often
with dense papillae along midrib and veins
.... *Erythrophyllum* (p. 118)

22. Midrib interrupted, darker than rest of plant
.... *Stenogramme*, in part (p. 126)

Midrib basal or absent 23

23. Midrib basal; sometimes seen only in older plants
.... 24

Midrib absent 26

24. Blades generally greenish-purple to pink; margins
smooth or with scattered proliferations ......
.... *Cryptopleura* (p. 143)

Plants cherry to brick red; blades highly branched
with dark veins in upper portions, or with numerous ruffles and/or proliferations on margins
.... 25

25. Macroscopic veins lacking in upper portions or
darker than rest of thallus ....
.... *Hymenena* (p. 145)

Macroscopic veins present in upper portions, not
darker than rest of thallus ....
.... *Botryoglossum* (p. 146)

26. Blades with numerous anastomosing macroscopic
veins .... *Polyneura* (p. 142)

Blades with microscopic veins only ....
.... *Acrosorium*, in part (p. 146)

27. Plants unbranched (may be split or have prolifera-
    tions) 28

    Plants branched 39

28. Entire plant hollow; filled with air, water, or mucus
    29

    Not hollow 30

29. Plants epiphytic on *Laurencia* (p. 154) ....
    .... *Erythrocystis* (p. 155)

    Not epiphytic .... *Halosaccion* (p. 136)

30. Plants soft to rubbery, wormlike ....
    .... *Nemalion* (p. 101)

    Plants bladelike 31

31. Blades epiphytic on sea grasses (p. 157) ....
    .... *Smithora* (p. 98)

    Blades otherwise 32

32. Blades saucer shaped and attached centrally to the
    stipe .... *Constantinea* (p. 103)

    Blades otherwise 33

33. Blades with papillae and/or proliferations ... 34

    Blades lacking papillae or proliferations 35

34. Texture coarse ...... *Gigartina*, in part (p. 128)

    Texture gelatinous ....
    .... *Grateloupia*, in part (p. 117)

35. Blades thin (one or two cells thick) ....
    .... *Porphyra* (p. 99)

    Blades thicker (many cells thick) 36

36. Blades yellowish-green to dark purple, rubbery,
    iridescent when submerged .. *Iridaea* (p. 133)

Blades otherwise 37

37. Blades rose red . . . . *Halymenia* (p. 115)
    Blades olive-purple to brownish-red 38

38. Blades olive-purple, soft, gelatinous . . . .
    . . . . *Grateloupia*, in part (p. 117)
    Blades brownish-red, slippery . . . .
    . . . . *Schizymenia* (p. 117)

39. Some branches hollow, filled with mucus 40
    All branches solid 41

40. Hollow parts large, terminal, grapelike . . . .
    . . . . *Botryocladia* (p. 136)
    Branchlets hollow and divided into series of bead-
    like cavities; branchlets much shorter but about
    the same diameter as solid branches . . . .
    . . . . *Gastroclonium* (p. 135)

41. Plants with series of circular or oval blades . . .
    . . . . *Opuntiella* (p. 124)
    Plants otherwise 42

42. Branching predominately dichotomous 43
    Branching palmate, distichous, or otherwise; not
    dichotomous 49

43. Branches round, less than 1 mm (0.04 in) in diam-
    eter with alternating light and dark bands
    . . . . *Ceramium* (p. 138)
    Branches flat 44

44. Branches covered with papillae . . . .
    . . . . *Gigartina*, in part (p. 128)
    Branches smooth or with scattered, small bumps,
    or large, wartlike outgrowths 45

45. Plants olive to purple; usually growing in bushy tufts to 15 cm (6 in) tall in mid intertidal zone .... *Rhodoglossum*, in part (p. 131)

Plants some shade of red 46

46. Branches stiff, erect; surfaces often with large, wartlike outgrowths; plants growing in dense tufts on sand-swept rocks .... *Gymnogongrus*, in part (p. 125)

Plants otherwise 47

47. Holdfast with spreading stolons .... *Rhodymenia* (p. 126)

Holdfast discoid 48

48. Blades thick, fleshy ... *Gracilaria*, in part (p. 121)

Blades thin, crisp .... *Stenogramme*, in part (p. 126)

49. All branches, or at least branchlets, distichous (in two rows along main axis) 50

Branching otherwise 58

50. Branchlets oppositely arranged along branches .... *Ptilota* (p. 140)

Branching alternate or irregular 51

51. Branching alternate 52

Branching irregular 54

52. Plants epiphytic ... *Microcladia*, in part (p. 139)

Plants not epiphytic 53

53. Ultimate branches about 5 mm (0.2 in) long, arranged in flat clusters around short laterals .... *Odonthalia* (p. 157)

Ultimate branches shorter, not arranged in clusters around short laterals . . . .
. . . . *Pterosiphonia*, in part (p. 151)
(If you got this far and your plant is not *Pterosiphonia*, you may have a *Laurencia* [p. 154], which is almost alternate, or a *Plocamium* [p. 124] with short laterals.)

54. Branch tips rounded with minute apical slit or depression; plants with acrid odor . . . .
. . . . *Laurencia*, in part (p. 154)
Branch tips rounded or pointed, but lacking apical slit or depression; odor not acrid 55

55. Lateral branches distinctly constricted at junction with main axis 56
Lateral branches not constricted 57

56. Branches less than 2 mm (0.08 in) broad; plants bushy . . . . *Pterocladia* (p. 101)
Branches more than 2 mm broad; plants with chlorine odor . . . . *Prionitis*, in part (p. 117)
(If branches are round, you may have an atypical *Neoagardhiella* [p. 120].)

57. Branchlets often dense, bending upward near junction with main axis (geniculate) . . . .
. . . . *Gelidium* (p. 101)
Branchlets sparse; branching especially distichous in terminal portions; not geniculate . . . .
. . . . *Gigartina*, in part (p. 128)

58. Branching mostly palmate or flabellate; plants generally fan shaped 59
Branching otherwise 61

59. Blades rose red, crisp, thin (one cell thick) . . . . . .
. . . . *Myriogramme* (p. 146)

Blades orange to dark red, greater than one cell thick, with or without bluish iridescence . . . . .
. . . . 60

60. Plants with bluish iridescence; blades slippery
. . . . *Fauchea* and *Fryeella* (p. 132)

Plants without bluish iridescence; blades coarse
. . . . *Callophyllis* (p. 120)
(If your plant is crisp, it may be *Rhodymenia* [p. 126], which is occasionally palmate.)

61. Plants composed of cylindrical main axis surrounded by "woolly" branches . . . .
. . . . *Callithamnion* (p. 139)

Branches not "woolly" 62

62. Plant parts cylindrical or only slightly flattened . .
. . . . 63

Plant parts distinctly flat 72

63. Plants bushy; branches covered with small (less than 0.5 mm [0.02 in]) spines . . . .
. . . . *Endocladia* (p. 114)

Plants bushy or otherwise; not covered with small spines 64

64. Branching pectinate throughout . . . .
. . . . *Microcladia*, in part (p. 139)

Branching not pectinate or pectinate only in branchlets 65

65. Branchlets pectinate, curving toward parent branch
. . . . *Plocamium*, in part (p. 124)

Branchlets not pectinate 66

66. Branchlets congested around and often obscuring main axis; plants wiry .. *Rhodomela* (p. 155)

   Branchlets and texture otherwise 67

67. Plants growing in high intertidal; soft, rubbery .... *Nemalion*, in part (p. 101)
   (If your high intertidal plant is rubbery but tough with many small branchlets, it is probably *Cumagloia* [p. 101].)

   Plants growing lower or in drift; not soft or rubbery 68

68. Tips of branches with apical pit, tuft of short (less than 2 mm [0.08 in]) hairs, or both 69

   Tips without pit or hairs 70

69. Branch tips with apical pit but lacking tuft of hairs; plants with acrid odor ....
   .... *Laurencia*, in part (p. 154)

   Branch tips with tuft of hairs growing from apical pit or pointed end .... *Chondria* (p. 151)

70. Branching mostly from base, sparingly branched above; branches generally less than 2 mm (0.08 in) in diameter; plants lax ....
   .... *Gracilaria* (p. 121)

   Branching mostly above base 71

71. Branches generally greater than 2 mm in diameter; plants stiff, turgid ... *Neoagardhiella* (p. 120)
   (One *Gelidium* [p. 101] has similar branching, but branches are wiry.)

   Branches to 2 mm in diameter; branching radial; plants bushy .... *Hypnea* (p. 125)

72. Blades delicate, thin (one cell thick), ribbonlike,

with microscopic veins and irregular marginal teeth .... *Acrosorium*, in part (p. 146)

Blades thicker, lacking marginal teeth 73

73. Blades compressed to slightly flattened, less than 4 cm (1.6 in) broad, covered with rod-shaped papillae .... *Gigartina*, in part (p. 128)

Blades lacking papillae 74

74. Main branches forming zigzag pattern; plants delicate with numerous small, curved branchlets .... *Plocamium*, in part (p. 124)

Branches not zigzag; plants coarse or slippery 75

75. Plants coarse with many congested, strap-shaped branches having small proliferations; generally growing in tide pools .... *Prionitis*, in part (p. 117)

Plants with a few broad, slippery blades arising from flattened, short stipe; blades with proliferations and spines .... *Grateloupia*, in part (p. 117)

3

The Green Algae (Chlorophyta)

The members of this phylum, which includes many fresh water forms, are easily recognized by their grass (dark) green color. Only rarely do species of other seaweed groups show a distinctly green color. Although most Chlorophyta are grass green, a few are very dark green, yellowish-green or even red orange (pl. 1*a*).

These algae are mostly small to moderate in size, only exceptionally exceeding 30 cm (1 ft) in greatest dimension. Some are finely branched, filamentous, or tufted plants; others are broad, membranous sheets (foliose). Most are intertidal species and a few, such as *Ulva* and *Enteromorpha*, are among the most frequently encountered algae in quiet bays, harbors, and boat landings. There are relatively few species of green algae in California, but some of them are very abundant.

Ulothrix (woolly hair)

Members of this genus are common in both fresh and salt water. Marine species generally occur as small tufts or extensive mats of unbranched filaments attached to rock or wood in the high intertidal zone. *Ulothrix pseudoflacca* (fig. 11*a*) occurs as free, unbranched filaments 0.5 cm (0.2 in) or less in length. The band-shaped plastid that encircles each cell is typical of the

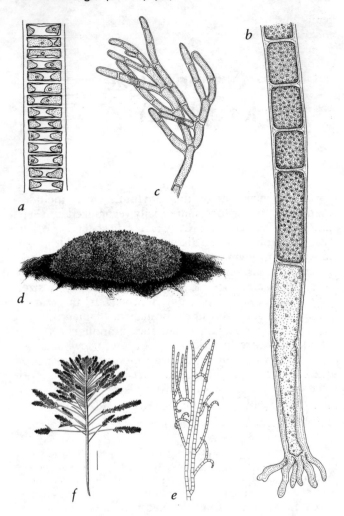

Fig. 11. *a, Ulothrix pseudoflacca*, portion of filament (0.08 mm, 0.003 in); *b, Chaetomorpha linum*, base of filament (1.4 mm, 0.06 in); *c, Cladophora columbiana*, branch (4 mm, 0.16 in); *d, C. columbiana*, entire plant (5 cm, 2 in); *e, Spongomorpha coalita*, branch (2 cm, 0.8 in); *f, Bryopsis corticulans* (7 cm, 2.8 in)

genus and can be used to distinguish it from *Rhizoclonium*. The latter, typically found in floating or attached masses in calm bays and marshes, is also small celled and filamentous, but the filaments are longer and more coarse and the cells have reticulate plastids.

Distribution: *U. pseudoflacca*, San Francisco to Los Angeles; other *Ulothrix* species, Oregon border to Los Angeles.

Chaetomorpha (hair shape)

Chaetomorpha linum (fig. 11*b*) is one of the more delicate forms of green algae in California. It resembles straight green or yellowish hair, sometimes white toward the ends of the filaments if spores or gametes have been released. Filaments are unbranched, usually between 5 and 30 cm (2–12 in) long, and frequently grow in groups of hundreds or thousands of individuals in sandy areas on rocks or around tide pools. Individual cells appear as tiny beads if a filament is held up to the light. *C. spiralis*, which occurs as a bluish-green epiphyte coiled around surf grass or other algae, has even larger cells, which can easily be seen with the naked eye. *Urospora*, a closely related alga that is common in the spring on the exposed coast, is similar in appearance to *C. linum*, but has very thin cross walls and produces anchoring rhizoids from cells near the base of the filament. Rhizoids arise only from the basal cell in *Chaetomorpha*.

Distribution: *C. linum*, Bodega Bay to San Diego; *C. spiralis*, Channel Islands to Mexican border; *Urospora* species, entire coast.

Cladophora (branch bearing)

This is a genus with numerous species, both in marine and fresh water environments. It is fairly readily recognized by its branched filaments in which the cross walls occur at rather close intervals and at the base of every branch. It lacks hooked branchlets. Most of the species are quite delicate forms, seldom reaching more than a few centimeters in length. Species can be difficult to determine, but one of the most common and distinctive in California is *C. columbiana* (fig. 11*c*, 11*d*), which grows at fairly high intertidal levels as a bright green, densely congested, spongy, hemispherical tuft (pl. 2*e*). This tuft is capable of holding sufficient water (together with sand) in the intricate network of its branches so that it can readily withstand long periods of exposure to air without drying out.

Another common, easy-to-recognize species is *C. graminea*, a dull green plant that grows in loose tufts made up of very stiff branches whose cells are often more than 1 cm (0.4 in) long. Plants are found in shaded areas from the mid intertidal zone to the subtidal zone and are frequently associated with sponges and tunicates.

Distribution: *C. columbiana*, entire coast; *C. graminea*, Santa Cruz to Los Angeles.

Spongomorpha (sponge shape)

Spongomorpha coalita, by far the most common species, is a finely and abundantly branched filamentous plant having the appearance of a piece of frayed green rope. It reaches 30 cm (1 ft) or more in length and may be encountered in a variety of intertidal and shallow subtidal habitats. The entangled, congested character of the branches results from the holding ability of multitudes of short, recurved branchlets (fig. 11*e*). The plant is otherwise much like a species of *Cladophora* (p. 49), consisting of branched, regularly septate filaments.

The ropelike strands are only part of the life history of this alga. The sporophyte is a small, unicellular green plant that lives in the tissue of the crustose red alga, "*Petrocelis*," a phase in some species of *Gigartina* (p. 128). There is some debate over the correct generic name for *Spongomorpha*; some phycologists call it *Acrosiphonia*.

Distribution: *S. coalita*, Oregon border to Point Conception.

Bryopsis (mosslike)

This genus is one of the few in our area with a thallus that is not divided into discrete cells. The branched axis, instead, consists of a tough, flexible cell wall that forms a hollow tube filled with a continuous matrix of multinucleate cytoplasm. Such construction is called siphonous (tubular) rather than filamentous. Cross walls only form where branches join the main axis when the branches become fertile.

Bryopsis corticulans is probably the most common species in California, forming dense, dark, soft clumps in areas of strong surf in the mid to low intertidal zone, as well as on pier pilings and boat docks in quiet waters. It is composed of numerous delicate, pinnately branched axes (fig. 11*f*). The green alga *Derbesia* and the yellow-brown alga *Vaucheria* (Chrysophyta) are also siphonous. But both are much smaller in diameter than *B. corticulans* and are not pinnately branched. *V. longicaulis* forms dark green, fine, hairlike mats on intertidal mud flats. *Derbesia* siphons are less than 0.07 mm in diameter and grow in lax tufts from 1 to 2 cm tall in the rocky intertidal and subtidal zones (also see p. 56).

Distribution: *B. corticulans*, entire coast; *V. longicaulis*, Bodega Bay to Monterey.

Enteromorpha (intestine shape)

Enteromorpha is closely related to *Ulva* (p. 52) and differs mainly in being hollow. Thus the hollow tube of *Enteromorpha* consists of a single layer of cells (fig. 12*c*), while the membrane of *Ulva* resembles a collapsed *Enteromorpha* in which the opposing walls have become adherent to each other (fig. 12*b*).

Enteromorpha is a cosmopolitan genus that may be encountered in almost any shallow water marine environment. It is especially prevalent on boat hulls, buoys, docks, and woodwork. It is commonly associated with *Ulva* in bays and marsh waterways (pl. 1*e*) and has considerable tolerance for fresh water.

One of the most common species is *E. intestinalis*, which is unbranched and arises solitarily from a small holdfast. It may be contorted and irregularly swollen, like a piece of intestine. Other common species such as *E. compressa* are branched, both at the base (fig. 12*c*) and above. Branching is variable and species are not easily distinguished. *Blidingia*, a closely related alga, is similar in appearance to *Enteromorpha* spp., but adheres to the substratum with a pad of cells rather than rhizoids. *B. minima* is commonly parasitized by a fungus that produces dark spots on the thallus.

Distribution: *E. intestinalis* and other *Enteromorpha* species, entire coast; *B. minima*, entire coast.

Ulva (marsh plant) Sea Lettuce

Their bright green color and abundance in a variety of habitats make the various species of *Ulva* among the most conspicuous of intertidal and shallow subtidal algae. Plants are especially common in bays, lagoons, harbors, and marshes (pl. 2*a*, 2*b*, 6*a*). They are also abundant in the open coast rocky intertidal zone, particularly in areas such as boulder fields that are regularly disturbed by sand abrasion or wave action. Rapid growth and production of vast quantities of spores and gametes are characteristics that make these plants true sea "weeds." Many species can use ammonia as a nitrogen source and are generally tolerant of organic and metal pollution. Thus, *Ulva* is often common in heavily polluted habitats.

The genus *Ulva* is easy to recognize by its thin, green, membranous thallus of only two layers of cells

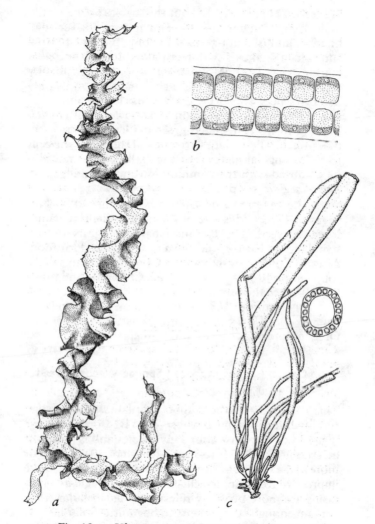

Fig. 12. *a, Ulva taeniata* (25 cm, 10 in); *b*, cross section of *U. taeniata* near blade margin (0.08 mm, 0.003 in); *c, Enteromorpha* sp., plant and cross section (plant, 6 cm, 2.4 in; cross section, 1 mm, 0.02 in)

(fig. 12a, 12b; pls. 2e, 6a), but the species are generally difficult to distinguish. *U. taeniata*, however, is very distinctive, and is characterized by long, narrow, twisted and ruffled segments with teeth along the lower lateral margins (fig. 12a). *Monostroma* spp. is occasionally found in quiet bays and has a similar form, but is only one cell thick. *Prasiola meridionalis*, found above the intertidal in the spray zone on rocks covered with guano (pl. 3a), forms convoluted green blades less than 1 cm (0.4 in) tall. These blades are one cell thick and the cells form rectangular patterns on the thallus. *Ulva* may also be confused with the common intertidal red alga, *Iridaea flaccida* (see p. 133).

The generic name *Ulva* was established by Linnaeus in 1753 and is one of the oldest scientific names for a seaweed. The common name suggests the resemblance to lettuce, and *Ulva* is widely used for food. After washing in fresh water, it can be added to salads, soups, seafood stews, and the like. Chopping and toasting improves the texture.

Distribution: *U. taeniata*, Oregon border to Point Dume; other *Ulva* species, entire coast; *P. meridionalis*, Oregon border to northern Channel Islands.

Codium (skin of an animal) Sponge Weed or Dead Man's Fingers

This is another of the siphonous plants, but is structurally different from *Bryopsis* (p. 51). *Codium fragile* (fig. 13a) is the most abundant species and is our most massive green alga. It reaches a length of 30 cm (1 ft) or more and a weight of over more than 0.5 kg (1 lb), quite impressive for a single cell! Large, multinucleate organisms are more properly referred to as acellular; they contain enough components to be multicellular but the components are not segregated by membranes or walls. *C. magnum*, a species found in Baja California, reaches lengths of more than 6 m (20 ft). The spongy texture results from an interwoven mass of fine, undivided si-

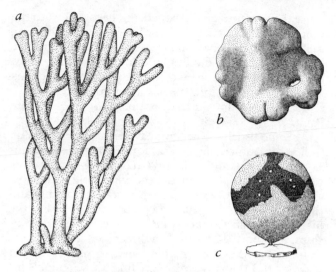

Fig. 13. *a*, *Codium fragile* (15 cm, 6 in); *b*, C. *setchellii* (7 cm, 2.7 in); *c*, "*Halicystis ovalis*" (1 cm, 0.4 in)

phons that end at the surface in closely packed, small, bladderlike swellings, each having (in some species) a sharp, microscopic point. Plants are deep green, cylindrical in form, and dichotomously branched. In addition to occurring as individual, drooping plants here and there on low intertidal rocks, C. *fragile* has been found at depths below 30 m (100 ft).

Encrusting, spongy, and almost black C. *setchellii* also occurs on our coast (fig. 13*b*). This species, like the brown seaweed *Laminaria sinclairii* (p. 84) and red *Gymnogongrus* spp. (p. 125), can withstand sand abrasion and burial for long periods and is often found on rocks where a sandy beach meets the rocky intertidal zone.

There are many other species of *Codium* found throughout the world. C. *fragile* subspecies *tomentosoides* is abundant along the Atlantic coast, where it commonly attaches to oyster shells. The added drag can cause the oyster to be dislodged and become lost to the fishery. This *Codium*, because of its high water-holding capacity, is used as packing material to ship live marine bait worms from New England to San Francisco Bay. Unfortunately, after the worms are sold, the live plants are often discarded into the bay and these discards have established an introduced population near Coyote Point.

Some usually unicellular blue-green or green, as well as other microalgae (p. 2) form symbiotic relationships with corals and other invertebrates. Many of these relationships are mutualistic: the alga living in the host tissue receives inorganic nutrients from the animal and the animal receives organic compounds made by the plant. Although *Codium* does not form such relationships, it is more than simply a direct source of food for some of the sea slugs (shelless marine molluscs) that graze on it. In the process of feeding, these animals ingest and somehow transfer intact chloroplasts from the alga into their own cells. The chloroplasts continue their photosynthetic processes and release compounds that are used by the host. The slugs thus become functionally plantlike.

Distribution: *C. fragile*, *C. setchellii*, entire coast.

"*Halicystis*" (sea bladder) Sea Bottle

Close observation of low intertidal and subtidal rocks encrusted with coralline algae will frequently reveal the presence of "*Halicystis ovalis*," a small, green sphere anchored by a single rhizoid (fig. 13c). These delicate little single-celled plants, usually most abundant in late spring and summer, are the gametophytic stage of *Derbesia marina* (p. 51). Prior to the discovery of this very heteromorphic life history, the two stages were classified separately. Since *Derbesia* was described first, it is now the correct generic name and "*Halicystis*," placed in quotes because it is no longer a valid species, is properly described as simply the gametophyte of *Derbesia*.

Although a small plant, the gametophyte is a very large cell and is frequently used by biologists in studies of how cells take up and use materials dissolved in the water around them.

Distribution: *D. marina* ("*H. ovalis*"), entire coast.

4

The Brown Algae (Phaeophyta)

The brown algae are almost exclusively marine plants that reach their greatest development in temperate and cold waters. California has one of the richest brown algal floras in the world, with almost eighty genera represented. Of these, the large brown algae called *kelps* are especially common and the state is famous for its submarine forests of giant kelp. In the northern hemisphere, these forests occur only along the west coast of North America, and harbor a diverse association of algae and animals; kelp forests are of great commercial and recreational importance.

Most of the brown algae are fairly large and some are the largest of all the algae. They generally have a distinctive habitat, branching, or shape, so recognition is relatively easy.

Giffordia (named for Miss Gifford)

There are several difficult to distinguish, small, filamentous brown algal genera. These frequently occur as epiphytes on older, often-deteriorating parts of other algae in the intertidal and subtidal zones. *Giffordia granulosa* (fig. 14a) is especially common in southern California in summer, where it appears as a much branched, delicate brownish fleece on rocks or other algae. *Giffordia* can be distinguished from the related *Ec-*

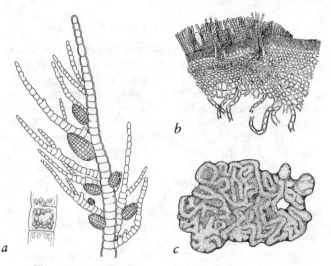

Fig. 14. *a, Giffordia granulosa,* branch (1.6 mm, 0.06 in) and cellular detail (0.16 mm, 0.006 in); *b, Ralfsia* sp., cross section (2 mm, 0.08 in); *c, Cylindrocarpus rugosus* (8 cm, 3.1 in)

tocarpus, since *Giffordia* has discoid plastids, and *Ectocarpus* has band-shaped plastids. Although commonly epiphytes, these genera, along with benthic diatoms (p. 3), are the weeds of the subtidal zone, usually the first plants to colonize newly exposed substrata.

The reproductive cells of these plants are produced in minute globular structures attached to the filaments (fig. 14*a*) and motile gametes or spores are released into the water. Given the great dilution and rapid water motion typical of the coastal ocean, one might expect that these and other marine plants would have evolved ways to increase the chances of free-swimming male and female gametes meeting after release. *Ectocarpus* uses a system analogous to that of many terrestrial insects: female gametes settle and secrete a chemical, ectocarpen, which attracts the males.

Distribution: *G. granulosa,* entire coast.

Ralfsia (after the British phycologist Ralfs)

In the mid and upper intertidal zone on rocks frequently exposed to sun and air is one of the seaweeds most tolerant of desiccation. *Ralfsia* occurs as a dark brown or blackish crust with slightly raised concentric ridges, often resembling the tar spots resulting from oil spills and submarine oil seeps. The alga is composed of short, erect filaments that are so densely packed and grown together that they form a firm, solid tissue (fig. 14*b*). The cells are extremely hygroscopic so the plants, although they become almost crisp dry in the sun, remain alive and continue growth with the first splash of the incoming tide. Crusts are firmly adherent and inflexible, but can be partly scraped or cut off with a sharp blade. They begin as more-or-less circular shapes, but by convergence and irregular growth, may form broad expanses up to 20 cm (8 in) or more across.

There are several crustose species related to *Ralfsia*, but they are poorly known and much work needs to be done on all aspects of their biology. Recent studies indicate that some of these "species" may be alternate phases in the life histories of other brown algae such as *Scytosiphon* (p. 63) and *Petalonia* (p. 65).

Distribution: *Ralfsia* spp., entire coast.

Cylindrocarpus (cylindrical fruit)

Another brown alga of prostrate habit and distinctive texture occurs on the same sun- and air-exposed rocks in the mid intertidal zone that support crusts of *Ralfsia*. *Cylindrocarpus rugosus* grows as a small, adherent cushion 2.5 to 5 cm (1–2 in) in diameter, of lobed and convoluted shape, and of a spongy, gelatinous consistency (fig. 14*c*). It spreads less extensively than *Ralfsia*, is much thicker, and can readily be removed from the rock with a dull blade. The substance is soft enough that a small bit can be crushed out on a slide to reveal the structure of branched filaments in a thick jelly.

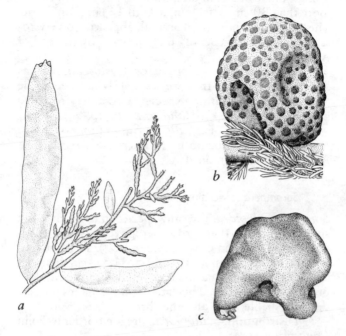

Fig. 15. *a*, *Coilodesme californica* on *Cystoseira* (12 cm, 4.7 in); *b*, *Soranthera ulvoidea* on *Rhodomela* (3 cm, 1.2 in); *c*, *Colpomenia peregrina* (5 cm, 2 in)

Distribution: *C. rugosus*, from Bodega Bay to Mexican border.

Coilodesme (hollowed tuft)

This genus is composed of hollow, generally compressed plants that grow as epiphytes on particular host plants. *Coilodesme californica* (fig. 15*a*), our most common species, is an obligate epiphyte on the upper, annual reproductive branches of *Cystoseira osmundacea* (p. 90). The light, olive-tan plants have the form of a delicate, flattened, elongate sack that may seem like a single blade but is, in fact, hollow. The thallus can be more than 30 cm (1 ft) long and to 12 cm (5 in) wide. The species is a short-lived annual, appearing on *Cystoseira* in May and generally disappearing by the end of August.

C. *plana* is also epiphytic on *Cystoseira*, but has linear, ruffled blades that are frequently tattered at the tips. C. *rigida*, a leathery species with a very rounded apex, is found only on *Halidrys dioica* (p. 92).

Distribution: C. *californica*, entire coast; C. *plana*, Cape Mendocino to Monterey; C. *rigida*, Los Angeles to Mexican border.

Soranthera (blooming mound)

Soranthera ulvoidea, the only species in the genus, is a summer annual that grows epiphytically on *Rhodomela* (p. 155) and *Odonthalia* (p. 157). The globular, hollow thallus is usually about 4 cm (1.5 in) in diameter and, as the name suggests, is covered with raised, dark brown sori (fig. 15 *b*). This latter feature makes it easily distinguishable from all other brown algae.

Distribution: S. *ulvoidea*, Oregon border to Point Conception.

Colpomenia (sinuous membrane)

Some of the most striking intertidal brown algae are the small, yellow-brown, hollow, bottlelike plants known as *Colpomenia*. These plants are widely distributed in warmer regions of the world and occur in a variety of forms. We find colonies on rocks (pl. 2*e*), in open places in the mid intertidal zone, or epiphytic on degenerating pieces of *Egregia* (p. 75) or other algae. The alga is usually only 3 to 5 cm (1.2–2 in) in diameter and, although generally hemispherical, may be smooth or bumpy. The wall of the hollow thallus is quite crisp and usually attached narrowly to the substratum. In Baja California where *Colpomenia* can be extremely abundant, walking over the intertidal zone is like stepping on popcorn.

A smooth species, *Colpomenia peregrina* (fig. 15*c*) is most abundant in northern and central California. *C. sinuosa* is a more southerly species that is often deeply convoluted and expanded. *Leathesia difformis* is a species of similar form and habit to *C. sinuosa*, but with a more slippery surface texture. *L. nana*, another globular brown, occurs as a small (to 0.5 cm or 0.2 in in diameter) generally solid epiphyte, especially on surf grass (p. 161). Microscopic examination of cross sections is needed to distinguish the two genera clearly.

Distribution: *C. peregrina*, Oregon border to San Diego; *C. sinuosa*, Santa Barbara to Mexican border; *L. difformis*, entire coast; *L. nana*, Oregon border to Santa Barbara.

Scytosiphon (leather tube)

This is a brown alga similar in form to the green *Enteromorpha intestinalis* (p. 52) and of similar size. Our most common species, *Scytosiphon lomentaria*, forms tubular thalli 5 to 15 cm (2–6 in) long, usually gregarious in clusters from a common, crustlike attachment. Small plants are slender and straight (pl. 5*d*). Large ones tend to be inflated and irregularly constricted. Although of worldwide distribution, the plant is quite strictly confined to the intertidal zone and usually occurs as the small, slender form in high intertidal pools. Less frequently, larger, constricted plants are found all the way down to levels exposed only by very low tides.

Distribution: *S. lomentaria*, entire coast.

Analipus (barefoot)

Analipus japonicus (fig. 16*a*) is the only species in our flora and occurs on mid to high intertidal zone rocks in areas exposed to heavy surf. It has a perennial, crustose holdfast and annual, upright branches that appear in spring and disappear in late fall. The erect branches, up

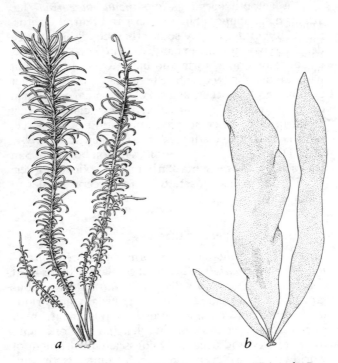

Fig. 16 *a*, *Analipus japonicus* (12 cm, 4.7 in); *b*, *Petalonia fascia* (12 cm, 4.7 in)

to 35 cm (14 in) tall, have a distinct central axis with numerous lateral branches. Externally, the plants resemble the red alga, *Cumagloia* (p. 101), but are distinctly brown and lack the latter's rubbery texture.

Analipus is commonly eaten in Japan, where it is called "matsumo." Branches are added fresh to soup or dried, salted, and cooked with soy sauce.

Distribution: *A. japonicus*, Oregon border to Point Conception.

Petalonia (leaflike)

Petalonia fascia is a widely distributed alga of the high intertidal zone, originally described from Atlantic Europe and known throughout much of the northern hemisphere. In California, it is usually found at levels of 0.6 to 1.2 m (2–4 ft) above mean lower low water on rocks, often subject to severe desiccation. Plants may also be found growing on surf grass (p. 161). It is a dark to greenish-brown, thin, and straplike plant 15 to 30 cm (6–12 in) long. Several blades arise from a very small, discoid holdfast (fig. 16b). They are smooth and entire, although sometimes undulate at the margins. The species is annual, appearing in the fall and disappearing the following summer.

Two other relatively small bladed browns are also found in California. *Endarachne binghamiae* is generally more golden brown than *Petalonia*, but grows in similar habitats in southern California, where *Petalonia* is rare or absent. *Phaeostrophion irregulare* grows in intertidal areas subject to sand burial and abrasion. Microscopic examination is necessary to distinguish the three genera accurately.

Distribution: *P. fascia*, entire coast (but rare in southern California); *E. binghamiae*, Point Conception to Mexican border; *P. irregulare*, Oregon border to Point Conception.

Pachydictyon (thick network)

The only intertidal brown algae of our region with smooth, thin blades and dichotomous to pinnate branching are *Pachydictyon coriaceum* (fig. 17a) and the closely related *Dictyota flabellata*. The two genera are distinguished only by the structure of the margins of the blades. *Dictyota* has only three tiers of cells throughout (fig. 17c), while *Pachydictyon* has five or more along the margins (fig. 17b). *P. coriaceum* is commonly deep brown, mostly 15 to 25 cm (6–10 in) long, smooth margined, and asymmetrically dichotomous. It inhabits southern California tide pools, and often is associated with the similar but smaller and paler *D. flabellata*.

Both species also occur in the subtidal zone, as does the generally larger *Dictyota binghamiae*. This latter species can be distinguished by the small teeth that commonly occur along the blade margins. The blades are also often full of holes produced by a small copepod (crustacean) that burrows into and eats the medullary and cortical cells from the inside. The outer cell walls of the cortex are left intact so the animal actually forms a cavity within the blade (less than 0.5 mm [0.02 in] thick!), which provides it with food and shelter.

Distribution: *P. coriaceum*, from Morro Bay south; *D. flabellata*, Santa Barbara to Mexican border; *D. binghamiae*, entire coast.

Zonaria (banded)

Zonaria farlowii (fig. 17d) is a handsome, olive-brown alga that ranges from sandy, low intertidal tide pools and rocks into the subtidal zone throughout southern California. Plants are densely bushy, 6 to 15 cm (2–6 in) tall and attached by a spongy, felted, fibrous holdfast. It is the only seaweed in which the tips of the growing blades appear as small fans with concentric lines. The alga is perennial and can survive for more than six months buried under sand.

The name, *Zonaria farlowii*, refers to the concentric zones or bands on the blades; it honors the late W. G. Farlow of Harvard University, the first resident American marine botanist.

Distribution: *Z. farlowii*, Santa Barbara to Mexican border.

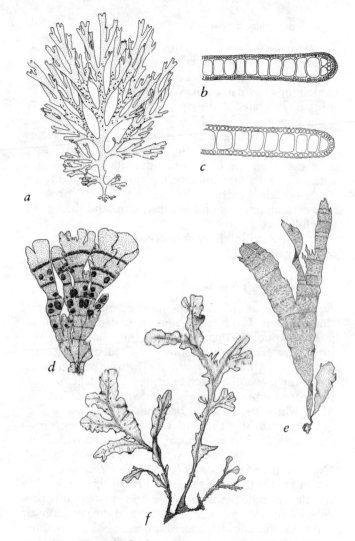

Fig. 17. *a*, *Pachydictyon coriaceum* (20 cm, 7.9 in); *b*, *P. coriaceum* cross section (0.3 mm, 0.01 in); *c*, *Dictyota* sp. cross section (0.3 mm, 0.01 in); *d*, *Zonaria farlowii* (6 cm, 2.4 in); *e*, *Taonia lennebackeriae* (15 cm, 6 in); *f*, *Dictyopteris undulata* (12 cm, 4.7 in)

Taonia (like a peacock)

Taonia lennebackerae is found in the same habitat as
Zonaria (p. 66) and also tends to be strap- or fan-
shaped overall. The blades are thin and generally light
to medium brown in color. When mature, they are often
split and lacerated at the tips, which gives them a more
linear appearance (fig. 17e).

Numerous algae can reproduce by vegetative as
well as sexual means. Both *Taonia* and *Zonaria* produce tissue on the basal rhizoids and blade margins
which develops into miniature plants called plantules.
These tiny duplicates of the parent break off, presumably attaching to a suitable substratum and then grow
into full-sized plants.

Distribution: *T. lennebackerae*, Point Conception
to Mexican border.

Dictyopteris (net wing)

Dictyopteris undulata is the most common species of
this southern California genus. It is distributed from
intertidal pools to depths of 18 m (60 ft) or more. Mature plants are densely, bushily branched, up to 30 cm (1
ft) high and attached by a coarse, fibrous base growing
into tough, woody older branches. The delicate, narrow, membranous blades have a distinctive midrib (fig.
17f), which persists and becomes part of the woody
branch system after the blade erodes away. Young, growing blades tend to be somewhat bluish and iridescent.

Dictyopteris, along with *Pachydictyon* (p. 65),
Dictyota (p. 65), *Zonaria* (p. 66), and *Taonia* (p. 68),
belongs to the primarily tropical brown algal order Dictyotales. Female plants produce eggs, male plants produce sperm with a single flagellum and the sporophytic
plants produce nonflagellated spores. Egg-, sperm-, and
spore-producing structures are usually aggregated in
sori that appear as light or dark spots on the thallus. In
some species, gametes are produced and released in a

cycle correlated with the tides. Such periodic mass release presumably increases the chances of fertilization.

Distribution: *D. undulata*, Santa Barbara to Mexican border.

Desmarestia (after the French naturalist, Desmarest)

The seaweed collector who mixes *Desmarestia* with other specimens will remember it well because of the damage that ensues, for the extraordinarily high sulfuric and malic acid content of the various kinds of *Desmarestia* cause marked discoloration of the pigments of other algae. It will even discolor and damage itself if a crushed or broken part is allowed to spread its acid cell sap on living surfaces. The normal function of the acid may be to prevent grazing. In the laboratory, sea urchins will eat *Desmarestia* when no other plants are available and when they do, the acid released erodes their calcareous "teeth." But we have observed sea urchins eating this alga in the field when other plants are available, so the acid may have other functions.

Desmarestia is generally a subtidal genus, but also occurs up into the low intertidal zone and is frequently found as drift in late summer. *Desmarestia ligulata* var. *ligulata* is our most common variety. Plants can grow more than 2 m (6.5 ft) long and are most abundant in summer at depths of 5 to 10 m (16–33 ft). The thallus is pinnately branched with a slender midvein and opposite branched veins, and may be finely branched and bushy or coarse with marginal spines (fig. 18). Mature plants commonly have a reddish coloration produced by the red alga *Acrochaetium desmarestiae*, whose tiny filaments may cover almost the entire blade. *D. ligulata* var. *firma* occurs in deeper water and, being broad and eroded at the tip when mature, resembles a tobacco leaf with midveins and opposite veins. *D. latifrons* has a narrow axis (less than 3 mm or 0.1 in wide) with mostly alternate branching. The relatively rare and difficult to distinguish *D. viridis* and *D. kurilensis* are strictly sub-tidal and have very fine, hairlike branches.

The plants described above are sporophytes; we know from culture studies that the gametophytes are all microscopic filaments. These tiny plants have been found only once in the field, growing in the tissue of a sea pen, a benthic animal related to sea anemones and jellyfish.

Distribution: *D. ligulata* (both varieties), Oregon border to San Diego; *D. latifrons*, Oregon border to Point Conception; *D. viridis*, entire coast; *D. kurilensis*, Oregon border to Monterey.

Nereocystis (sea nymph bladder) Bull Kelp

Nereocystis luetkeana is the most massive kelp of northern California. It grows up from deep water as an enormously long, hollow, hoselike stipe (up to 35 m or 115 ft), gradually enlarging at the top to a huge, elongate air chamber and terminal bulb (pl. 5*a*). This massive float supports 32 to 64 blades that may be 3 m (11 ft) long and hang down in quiet water. The upper stipe and bulb of older plants are often covered with epiphytes and one species of *Porphyra* (p. 99) is found nowhere else. Submarine groves of these giant algae show on the surface as a great many floating bulbs about the size of a sea otter's head. Indeed, it is often among the bull kelp that sea otters rest and feed, and one must look sharply to distinguish the two from a distance.

After a storm, bull kelp may be found littering beaches, their small holdfasts torn loose from the sea floor. Despite their size, these huge plants usually complete their growth (more than 25 cm or 10 in per day), mature, produce spores, and are cast up, all within a single year. Spores are formed in sori on the blades, and during the reproductive season the entire sorus is released from the blade. The sorus sinks to the bottom and releases its spores, presumably to insure that a

Fig. 18. *Desmarestia ligulata* var. *ligulata* (15 cm, 6 in)

maximum number of spores will reach a suitable substratum.

Pacific Indians prepared numerous dishes from *Nereocystis*. The blades can be washed, dried, and eaten like potato chips or added to other foods and the stipes can be prepared as sweet pickles. In addition to food uses, the hollow thallus can be made into musical instruments and dolls. Perhaps the easiest instrument to make is a trumpet, produced by cutting off the upper half of the bulb and lower part of the stipe. Entire bands, with each member playing some instrument made from bull kelp, have performed at marine laboratories. To make a doll, the blades ("hair") are trimmed as desired and various features are carved into the air chamber and/or bulb while the alga is wet. Subsequent drying in a warm room not only preserves the figure but, as the plant shrinks in the process, produces interesting variations on the original carving.

Distribution: *N. luetkeana*, Oregon border to Pismo Beach.

Pelagophycus (open ocean seaweed) Elk Kelp

This majestic alga is second only to giant kelp (p. 73) in size. *Pelagophycus porra* grows on the outer edges of giant kelp forests and arises from depths of 27 m (90 ft) or more (pl. 5 c). The very long, single stipe may extend to the surface from a fist-sized holdfast. The top of the stipe enlarges to form a hollow bulb more than 15 cm (6 in) in diameter, and this bears a pair of antlerlike branches from which huge, broad, membranous blades hang down or stream out like brown curtains (fig. 19) parallel to the surface in strong currents. The slender stipe is often broken by storms and the blades eroded away leaving the huge, antlered bulb floating far out to sea. Plants are known to drift at least 480 km (300 mi). Spanish mariners long ago were familiar with these floating objects, which they called "porra," and used them as an indication of their approach to the Mexican coast (California) on the long voyages from the Philippines. An account of the "Manila Ship" in Anson's Voyages (1748) reads as follows:

> and when she was run into the longitude of 96° from Cape Espiritu Santo, she generally meets with a plant floating in the sea, which being called *Porra* by the Spaniards is, I presume, a species of sea-leek. On sight of this plant, they esteem themselves sufficiently near the California shore, and immediately stand to the southward; and so much do they rely on this circumstance that on the first discovery of the plant, the whole ship's company chant a solemn *Te Deum*, esteeming the difficulties and hazards of their passage to be now at an end; and they constantly correct their longitude thereby, without ever coming within sight of land.

Distribution: *P. porra*, Point Conception to Mexican border.

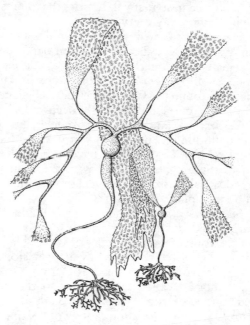

Fig. 19. *Pelagophycus porra*, two young plants (left, 2 m, 6.6 ft; right, 50 cm, 20 in)

Macrocystis (large bladder) Giant Kelp

The Pacific coast giant kelp forests (fig. 3; pl. 3*d*) are famous the world over. Not only are they the most extensive and elaborate submarine forests of the world, containing a very diverse array of plants and animals, but they also provide the raw material for a large industry. The harvest of giant kelp in California is regulated by the Department of Fish and Game and areas must be leased for commercial cutting. Millions of dollars have been spent on attempts to manage and restore giant kelp forests. Algin, the principal product extracted from the plant, is utilized in so many industries that almost all of us use it in some way every day.

Macrocystis is widespread in cool water, occurring in Australia, New Zealand, Argentina, Chile, Peru, and Pacific North America, but it is best developed as the species *M. pyrifera* along northwestern Baja California and around the California Channel Islands. The densest forest measured formerly occurred along the Palos Verdes headland near Los Angeles, but metropolitan pollution and unusually warm water destroyed it completely by 1959. Plants have recently begun to return to the area, coincident with the reduction of toxic industrial wastes in the local sewage effluent and aided by transplantation efforts.

Macrocystis is well known for its bladder-based, brown, wrinkled blades commonly found along the shore, and for its massive tangles of stipes, blades, and holdfasts that litter the beaches after a storm. The plants grow mainly in depths of 6 to 18 m (20–60 ft) and in groups are called a forest, because the erect, entwined bundles of fronds (stipes and blades) resemble a tree trunk, and the spreading canopy of floating "stems and leaves" the crown of a tree (pls. 3*d*, 4*c*). *Macrocystis* anchors to rocks or occasionally in sand by a huge holdfast of rootlike haptera, and new fronds, composed of stipes and attached blades, grow up to the surface at extraordinarily rapid rates. Growth in length of individual fronds can be more than 35 cm (14 in) per day, the fastest in the plant kingdom, exceeding that of fast-growing tropical bamboos.

The massive giant kelp plants are all sporophytes. They produce motile spores from sori on special blades near the holdfast. These spores develop into microscopic, sexual plants that rarely grow large enough to be seen with the naked eye (pl. 4*b*). These delicate male and female gametophytes produce sperm and eggs that, upon fertilization, recreate the large sporophyte generation (fig. 9*a*). Circumstances suitable for the growth of these minute sexual plants are critical to the survival of kelp forests.

The California kelp forests were first exploited commercially during World War I as a fertilizer resource in the absence of German potash. In the early 1930s, the algin extraction industry was developed and algin extraction is now by far the most important use of *Macrocystis* in this country. Algin is a hydrophilic colloidal substance very effective as an emulsifying and suspending agent. As such, it is extensively used in preparing many processed foods and numerous other products such as paints, cosmetics, pharmaceuticals, and sizings.

Giant kelp harvesting is carried out mechanically by means of a ship designed with a mowing and hauling device by which the kelp is cut no more than 1.2 m (4 ft) below the surface and drawn up into the vessel for transport to the processing plant.

Macrocystis integrifolia, a primarily intertidal species, also occurs along our rocky shores. It is distinguished by a flattened, creeping, rhizomelike holdfast with haptera along the lateral margins.

Distribution: *M. pyrifera*, entire coast; *M. integrifolia*, Oregon border to Point Conception.

Egregia (remarkably great) Feather Boa Kelp

Another large kelp that can be recognized at a glance is *Egregia menziesii*, which is unique in our area, having long, flat axes bearing numerous flat, feathery lateral blades and basal floats along their whole length (fig. 20a, pl. 3e). This is a kelp of shallow water, usually of surf-swept areas, and is one of our most abundant large algae in the low intertidal zone and at depths to 6 m (20 ft). During fall and winter, one often finds numerous juvenile and developmental stages of *Egregia* that appear so different from the adults that they may be mistaken for a different species. Plants less than 0.3 m (1 ft)

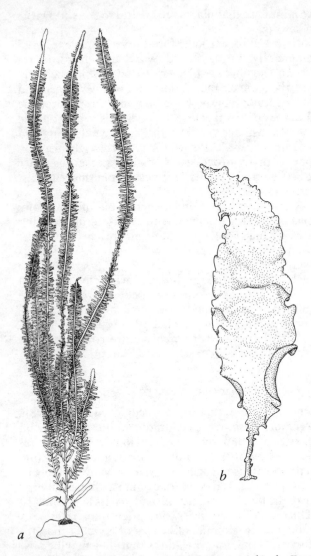

Fig. 20. *a*, *Egregia menziesii* (5 m, 16 ft); *b*, *E. menziesii*, young plant (15 cm, 6 in)

long show a broad terminal blade from a short stipe that may or may not exhibit the beginning of lateral branches or of blades as the stipe elongates and flattens (fig. 20*b*). Growth occurs in an intermediate zone at the base of the primary blade, which gradually erodes away as the stipe elongates.

During winter and spring, lateral blades of young and mature vegetative plants are characteristically leaf-like and elongate. In many old plants during summer, however, the blades become finely dissected and hair-like, so that fragments of the two growth forms appear very different. The flat axis and holdfast frequently appear eroded and full of holes. This damage is caused by the limpet (snail) *Notoucmea insessa*, which grazes almost exclusively on *Egregia*.

Distribution: *E. menziesii*, entire coast.

Postelsia (after the Russian naturalist, Postels) Sea Palm

Were we to elect a California state seaweed, this would be a fitting candidate, for it is perhaps the most distinctive and striking of our intertidal algae. It has the remarkable form of a palm tree, to about 60 cm (2 ft) tall, with a substantial holdfast, tough, hollow, rubbery stipe, resilient to the most terrible pounding of the surf, and a terminal cluster of slender, drooping blades that withstand the waves as a coconut palm survives a hurricane (pls. 4*a*, 12*d*). *Postelsia palmaeformis*, the only species, is an annual, and in February, the groves of juvenile, yellow-brown sea palms may be seen springing up to reoccupy their habitat among the perennial *Lessoniopsis* (p. 78) and mussels in the mid intertidal zone. Young sea palms grow up in areas of bare substratum and their holdfasts overgrow other algae and sessile animals that settle nearby. These associated organisms are torn loose along with *Postelsia* during fall and winter storms, providing new, uncolonized space for the next generation of gametophytes and sporophytes.

Sea palms are considered a delicacy by many. Not only can the stipes be made into sweet pickles, but the stipe and blades, after washing in fresh water and steaming, are eaten with lemon juice and soy sauce. *Postelsia* harvesting could have serious consequences for local populations, however, as we have seen entire groves of nothing but cut stipes; every plant had its top removed. The spores required to reproduce the next generation are produced in the blades and spore dispersal is known to be limited. Total blade removal in a grove prior to spore production could virtually eliminate the next generation and allow other organisms to take over areas formerly inhabited by sea palms. A similar problem could occur if annual bull kelp (p. 70) were harvested for algin extraction. Bull kelp and giant kelp plants often grow close together in central California, and the California Department of Fish and Game requires that giant kelp harvesters take care not to have more than 5 percent bull kelp in their harvests.

Distribution: *P. palmaeformis*, Oregon border to Morro Bay.

Lessoniopsis (after the French naturalist, Lesson)

One of the locally abundant kelps along the surf-swept rocky shores of California is *Lessoniopsis littoralis*, which is very well adapted to withstand the most violent agitation of the sea (pl. 2*d*). It consists of a coarse, conical, woody base attached by strong and compact haptera. The lower stipe is repeatedly dichotomous and each upper branch terminates in a long, narrow, flat blade with a strengthening midrib (fig. 21*a*). The whole hank of blades lashes back and forth with each wave, keeping the substratum around each plant clear of most other organisms. Unlike *Postelsia* (p. 77), which lives in the same inhospitable environment, *Lessoniopsis* is perennial and may live for many years to become quite gnarled.

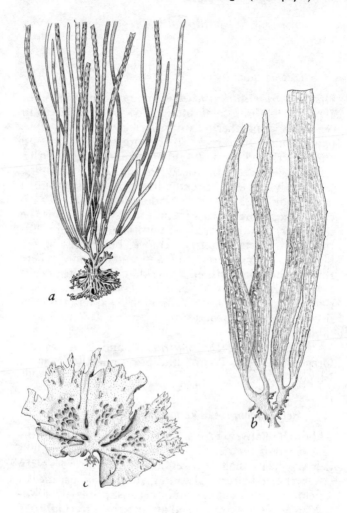

Fig. 21. *a, Lessoniopsis littoralis* (40 cm, 16 in); *b, Dictyoneurum californicum* (30 cm, 12 in); *c, Hedophyllum sessile* (30 cm, 12 in)

Distribution: *L. littoralis*, Oregon border to Big Sur.

Dictyoneurum (nerve net)

Dictyoneurum californicum is a common inhabitant of exposed, low intertidal and shallow subtidal areas in central and northern California, growing in dense patches attached to rocks. Blades are generally around 1 m (3.3 ft) long, 6 cm (2.4 in) broad and covered with a rectangular pattern of raised ribs (fig. 21*b*). Although apparently growing directly from the substratum, a closer look reveals that the blades are attached to a dichotomously branched stipe, most of which forms the tenacious, prostrate holdfast with lateral haptera. Entire plants are difficult to remove, but pieces of the blades commonly break off and come ashore where they can be distinguished from other fragments of drift kelp by the pattern of the ribs. *Dictyoneuropsis reticulata*, a related but less common alga, has a similar overall form but broader blades with a conspicuous midrib.

Distribution: *D. californicum* and *D. reticulata*, Oregon border to Point Conception and the Channel Islands.

Hedophyllum (seatlike leaf)

Although relatively rare at its southern limit in California, *Hedophyllum sessile* is one of the most abundant kelps in the Pacific northwest and Alaska where it often covers the mid intertidal zone, giving the appearance of a field of brown cabbage. Numerous, primarily subtidal, algal species can extend up into the intertidal zone under the blades of *Hedophyllum*, which provide protection from direct sunlight and drying. The plant is distinguished from other large, brown algae by the absence of a stipe; haptera grow directly from the lower margin of the large, coarse, undivided blade (fig. 21*c*).

Blades in quiet water are highly corrugated and up to 80 cm (31 in) wide. In more exposed habitats, they are linear and smooth.

Distribution: *H. sessile*, Oregon border to Big Sur.

Pterygophora (wing bearing)

Pterygophora californica is a large, long-stiped alga that lives from the low intertidal zone to depths of more than 24 m (80 ft). The stiff, erect stipes can be up to 2 m (6.6 ft) long and bear several pinnately arranged sporophylls in the upper portion. The single terminal blade has a broad, thickened stripe down its middle (fig. 4).

Pterygophora, like *Eisenia* (p. 83) and some species of *Laminaria* (p. 84), develops growth rings in the stipe cortex somewhat like the rings in terrestrial trees. These rings appear to be annual and we have counted more than sixteen in the stipes of very large plants from Carmel Bay. These long, "woody" stipes often accumulate in the drift and, when dried rapidly, have been used as a substitute for wood in making fences.

Dense groves of *Pterygophora* may form a complete understory canopy beneath the surface cover of *Macrocystis* (fig. 3) in the subtidal zone. Diving close to the bottom beneath these combined canopies is like walking into a theater while the movie is in progress; almost total darkness prevails until your eyes adjust. This severe light reduction inhibits the growth of low-lying red algae, which are more abundant outside the groves, an effect similar to the reduction of understory vegetation under some dense terrestrial forest canopies.

Distribution: *P. californica*, entire coast.

Alaria (wing)

Kelps in this genus are most common in the far north Pacific regions. They are best developed in extensive submarine beds in Alaskan and Siberian waters, where

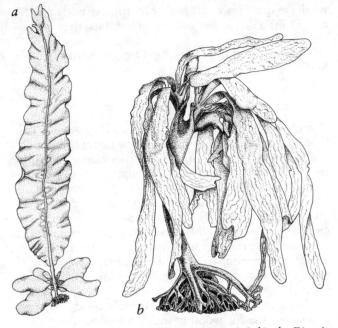

Fig. 22. *a*, *Alaria marginata* (1 m, 3.3 ft); *b*, *Eisenia arborea* (60 cm, 24 in)

the primary blades may exceed 25 m (82 ft) in length and produce an inflated, floating central midrib. Our common species, *Alaria marginata* (fig. 22*a*), is one of the most southerly and does not float. It is often abundant at the lowest intertidal levels and is easily recognized by its short stipe (3–8 cm [1–3 in]), large, ruffled primary blade with prominent midrib and by the several much smaller and shorter sporophylls arising below the main blade. Plants growing in areas more exposed to wave action have short, tattered primary blades.

Distribution: *A. marginata*, Oregon border to Pismo Beach.

Eisenia (after the naturalist, Gustav Eisen)

The famous submarine gardens of Santa Catalina Island contain some of the finest stands of *Eisenia arborea* and have given innumerable visitors on the glass-bottom-boat excursions a memorable impression of marine vegetation. *Eisenia* is our most treelike kelp (pl. 3c), for it has an erect, woody stipe up to 1 m (3.3 ft) long supporting a pair of branches from which the leafy blades hang. Smaller, mature plants may be found intertidally in several of our rocky areas (fig. 22b) and juvenile specimens, which may lack conditions suitable for maturation, are often abundant. These, like young *Egregia* (fig. 20b) and *Macrocystis* plants, are unlike adults in consisting at first only of a short stipe and a broad primary blade. The young *Eisenia* blade erodes away, while growth proceeds from either margin of its base. As lateral blades grow out from these basal margins, which elongate and thicken, they form a pair of prong-like false branches that support the blades. This two-clumped feature renders it markedly distinct from the sea palm (p. 77) of surfy northern California shores.

Eisenia is particularly well adapted to shallow subtidal areas exposed to strong surf. The branches orient perpendicularly to the surge, and the tenacious holdfast and rubbery stipe make the plant difficult to dislodge or break. Moreover, the blades clump together and stream out parallel to the bottom in strong surge, reducing the drag on the plant.

Eisenia arborea is commonly called the southern sea palm because of its palmlike form and because it was once thought to occur only south of Point Conception. Overlooked old records and recent collections indicate it occurs in northern British Columbia, and is actually found further north than *Postelsia*, the sea palm.

Distribution: *E. arborea*, vicinity of Carmel Bay and from Point Conception to Mexican border.

Laminaria (thin plate) Oar Weed

This is a genus of considerable historical, economic, and ecological significance among the algae. The oar weeds were some of the most important seaweeds long ago collected from drift for burning down into soda ash that was used for making soap and glass. The burned ashes of these seaweeds were originally called "kelp" in Europe and, in more recent times, the word has been used as the common name for large brown algae. *Laminaria* and related genera are much more common in the temperate, coastal waters of the world than surface canopy formers such as *Macrocystis* (p. 73). Areas where the former cover the bottom with their broad blades are called kelp beds, while we refer to areas with giant and bull kelp as kelp forests to emphasize their greater vertical structure.

Adult *Laminaria* is readily recognized by its single, unbranched stipe bearing a single blade that may be torn or split, but is not branched and does not have veins or ribs (fig. 23). Plants may be quite large, reaching over 5 m (16.4 ft) in length. *L. dentigera* is one of these large species with a stiff stipe generally about 70 cm (2.3 ft) tall and a broad, smooth blade that develops numerous longitudinal splits in shallow, surf-swept areas (fig. 23a; pl. 3b). It grows in beds in the low intertidal and shallow subtidal zone and is of striking appearance on an afternoon low tide when the sun shines through the rich, brown blades and reflects from the darker stipes with each slacking of the water after a wave. *L. farlowii*, almost exclusively a subtidal species, grows at depths to 50 m (165 ft) in southern California. Its short stipe and broad, long, rumpled blade are distinctive (fig. 23b). Blades frequently have a reddish caste produced by small red algae growing in the tissue. A diver can easily become seasick watching these blades roll back and forth along the bottom with the surge. *L. sinclairii* is a smaller species consisting of a branched, prostrate stolon from which dense clumps of narrow-bladed stipes about 30 cm (1 ft) high arise (fig. 23c). It is particularly common in areas periodically covered by sand. *L. ephemera* is a relatively rare species similar in form to *L. sinclairii* but with a discoid holdfast. Juvenile plants of all species except *L. sinclairii* are very difficult to distinguish, both from each other and from juveniles of related genera in the order Laminariales (*Macrocystis*, *Egregia*, etc.). The short-stiped, thin-bladed juvenile form may represent the optimal adaptation to water motion close to the bottom. Plants begin to develop their adult form when they are between 5 and 15 cm (2–6 in) long.

Oar weeds have long been harvested for their iodine after "kelp" production became uneconomical, and in more recent decades, they have become important sources of algin in Atlantic and eastern Pacific waters. Several species are common food seaweeds in the orient (Japanese *kombu* and Chinese *haidai*). They are important fertilizer sources in many areas and are widely used along European coasts as stock feed. The Peoples Republic of China has recently begun large-scale cultivation of *Laminaria* for a variety of uses. Using modern aquacultural techniques of raft cultivation, artificial fertilization, and genetic selection, their harvest of cultivated *Laminaria* is ten times greater than the harvest of *Macrocystis* from natural forests in California, and cultivated areas occupy some 10,000 hectares (24,700 acres) of sea surface. *Laminaria* farming for food has also recently started in Canada.

Distribution: *L. dentigera* and *L. farlowii*, entire coast; *L. sinclairii*, Oregon border to Santa Barbara; *L. ephemera*, Oregon border to Big Sur.

Costaria (riblike)

Costaria costata (fig. 23d) is a striking brown alga, much like *Laminaria* (p. 84) in form, but provided with five prominent longitudinal ribs in the blade, two raised on one side and three on the other. Between the ribs, the blade is strongly wrinkled. It is another common alga of the lowermost tide levels on surf-swept shores. Two other less common kelps with single, ribbed blades, *Agarum fimbriatum* and *Pleurophycus gardneri*, also occur in California. *Agarum*, found in deep water in southern California, has a broad midrib and the rest of the blade is very rumpled and full of holes. *Pleurophycus* also has a broad midrib, but the blade lacks holes.

Distribution: *C. costata*, Oregon border to Los Angeles; *A. fimbriatum*, Point Conception to Mexican border; *P. gardneri*, Oregon border to Bodega Bay.

Fucus (seaweed) Rock Weed

The ancient Latin name *fucus* was originally applied to seaweeds in general. In the 1750s, *Fucus*, *Ulva*, and *Conferva* were used as the first generic designations for most of the algae. As currently described, the genus *Fucus* still includes some of the most common intertidal plants in the temperate coastal waters of the northern hemisphere and, along with a few related genera that are commonly called rock weeds, fucoids or bladderwracks dominate the rocky shores of the North Atlantic.

In California, the highly variable mid to upper intertidal zone species is *Fucus distichus*. These dichotomously branched, olive-green to dark brown plants can be up to 25 cm (10 in) long and have a prominent midrib in older portions of the blades (pl. 5b). The tips of the blades become swollen and covered with small bumps when reproductive. *Fucus* and related genera (*Pelvetia*, *Cystoseira*, and others) do not produce spores or gametophytes, thus having an animallike life history (fig. 9b).

The swollen tips, called receptacles, may represent up to half of the length of the thallus and often become hollow and inflated in quiet water, floating the plant off the substratum at high tide. In areas exposed to moderate wave action, the blades lash about over the

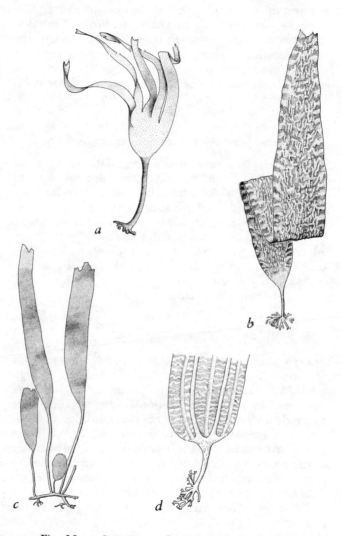

Fig. 23. *a, Laminaria dentigera* (60 cm, 24 in); *b, L. farlowii* (70 cm, 28 in); *c, L. sinclairii* (30 cm, 12 in); *d, Costaria costata*, lower portion of plant (15 cm, 6 in)

substratum and may prevent other organisms from living beneath them. In calmer habitats during low tides, however, the moist, shaded environment beneath the blades provides a refuge from desiccation for a variety of plants and animals.

Distribution: *F. distichus*, Oregon border to Point Conception.

Pelvetia (after the French botanist, Pelvet)

The high, exposed rocks in the intertidal zone are commonly draped with an olive-greenish-brown plant about 30 cm (1 ft) long (pl. 5*b*). This is *Pelvetia fastigiata*, which is adapted to a life of alternate submersion in sea water and long exposure to air. The thick, narrow branches are dichotomous and, like *Fucus* (p. 86), with which it often grows, mature, fertile plants have swollen branch tips from which the reproductive cells are extruded from numerous pores.

Distribution: *P. fastigiata*, entire coast.

Pelvetiopsis (diminutive of Pelvetia)

At even higher levels than *Pelvetia* (p. 88) and *Fucus* (p. 86), indeed, on rocks scarcely submerged even at high tide, is a densely clustered, small, light-tan colored alga 3 to 8 cm (1–3 in) tall, *Pelvetiopsis limitata*. It consists of closely dichotomous branches that tend, especially in younger stages, to be arched inwardly (fig. 24*a*). Mature plants have swollen, fertile branch tips and the presence of these tips on the relatively small *P. limitata* can be used to distinguish it from juvenile *Pelvetia* and *Fucus*. *Pelvetiopsis* spends more of its life in the air than in water and may be observed at almost any state of the tide but the highest.

Distribution: *P. limitata*, Oregon border to Morro Bay.

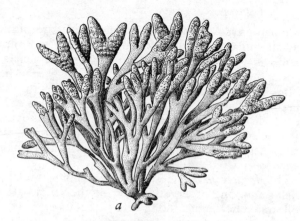

Fig. 24. *a, Pelvetiopsis limitata* (7 cm, 2.8 in); *b, Hesperophycus harveyanus* (25 cm, 10 in)

Hesperophycus (western alga)

Hesperophycus harveyanus, endemic to California and northern Pacific Mexico, resembles *Fucus* (p. 86) and replaces it in southern California. Plants are generally about 30 cm (1 ft) long with dichotomously branched, often undulate blades and terminal receptacles that may become inflated as in *Fucus* (fig. 24*b*). The alga can be easily distinguished from *Fucus* by the tiny tufts of hairs that occur in two parallel rows on either side of the midrib. These hairs are especially evident when the thallus becomes partially dried during a low tide.

Hesperophycus is especially common on the Channel Islands and observations during the Santa Barbara oil spill indicate that, along with surf grass (p. 161), it is particularly susceptible to damage by oil pollution.

Distribution: *H. harveyanus*, Santa Cruz to Mexican border.

Cystoseira (bladder chain)

Three species of *Cystoseira* occur in California, of which *C. osmundacea* is by far the most common. When reproductive, it is a large, bushy, finely branched brown alga from 1 to 10 m (3 – 33 ft) long that grows in large tide pools and from the low intertidal zone to depths of more than 10 m. Plants have a perennial, dark brown basal portion consisting of a gnarled stipe surrounded by radially arranged branches (fig. 25*a*). These pinnate branches with central veins resemble the leaves of *Osmunda* ferns and the specific name emphasizes this similarity. In spring and summer, the lower branches produce slender, highly branched upper branches with long, cylindrical axes. The ends of these upper portions develop a distinctive, beadlike series of 5 to 12 small, spherical floats with short reproductive branches at their ends. This mass of buoyant material frequently forms a dense surface canopy between deeper-growing *Macrocystis pyrifera* and the shore (fig. 3). Several other algae grow on the upper branches,

Fig. 25. *a, Cystoseira osmundacea* (60 cm, 24 in); *b, C. osmundacea*, juvenile (4 cm, 1.6 in); *c, Halidrys dioica*, flattened float (2.5 cm, 1 in)

but especially *Coilodesme californica* (p. 61). After the first fall storms, the upper branches drift ashore in brown, spaghettilike masses.

Juvenile *Cystoseira* have a distinctive appearance, lacking upper branches and having small, entire lower branches radially arranged around the stipe (fig. 25 *b*). Juveniles are commonly observed growing deeper than 10 m, but probably because of the reduced light, they rarely develop further.

Distribution: *C. osmundacea*, entire coast.

Halidrys (sea tree)

Juvenile and vegetative plants of *Halidrys dioica* cannot be distinguished from juveniles of *Cystoseira* (p. 90). As soon as the upper portions with floats appear, however, they may be recognized at a glance for the series of floats on *Halidrys* are flattened instead of spherical and beadlike (fig. 25 *c*). *Halidrys*, like *Cystoseira*, is a common plant of lower intertidal tide pools and shallow subtidal waters, where it may reach 2 m (6.5 ft) in length. Plants have been growing along southern California for a long time, for in the Doheny Palisades near San Juan Capistrano and in other Miocene (13 to 25 million years old) diatomite and siltstone shales of this region, fossil specimens are common and show only minor variations from present-day plants. Although *Halidrys* is old and may once have been widespread, it survives today only in California, Mexico, and Atlantic Europe.

Distribution: *H. dioica*, Los Angeles to Mexican border.

Sargassum (from Portuguese *sarga*, a kind of grape)

Portuguese mariners of the fifteenth century named that vast Atlantic eddy filled with drifting seaweed buoyed up by their small, grape-shaped air vesicles the Sargasso Sea. We now give these plants the name

Fig. 26. a, *Sargassum muticum* (40 cm, 16 in); b, *S. palmeri*, blade and float (3 cm, 1.2 in); c, *S. agardhianum*, blade and float (2 cm, 0.8 in)

Sargassum, representing a genus of brown algae with a wide tropical and subtropical distribution. There are three low intertidal−shallow subtidal zone species in California, two native in the south and one introduced which occurs throughout the state. *S. palmeri*, a native, is especially common at Catalina Island. It approaches 1 m (3.3 ft) and has much-dissected, narrow blades (fig. 26b). *S. agardhianum*, the other native, is more common on the mainland. It is of similar size, but has clusters of undivided, leaflike blades (fig. 26c).

The large (to 10 m [33 ft]) introduced *S. muticum* has undivided, leaflike blades that occur singly along the thallus (fig. 26*a*).

S. muticum was probably introduced on Japanese oysters imported to the Puget Sound region of Washington in the 1930s. It was reported in northern California in 1963 and has since spread to San Francisco Bay, Catalina Island, the mainland of southern California, and into Baja California. Drift plants have also been found in Monterey. The plants are generally found in sheltered bays and harbors; experiments in San Diego have shown that they are poor competitors for space with native seaweeds of the open coast. *S. muticum* was recently introduced to France and England, and British phycologists have mounted an extensive eradication program that has not been successful.

Distribution: *S. palmeri*, Catalina Island to Mexican border; *S. agardhianum*, Point Dume to Mexican border; *S. muticum*, entire coast.

5

The Red Algae
(Rhodophyta)

The red algae are extremely common seaweeds of worldwide distribution. There are more than 4,000 species and they occur in almost every conceivable benthic habitat in the sea, from the highest intertidal levels to depths where plant growth is limited owing to lack of light. Only a few live in fresh or brackish water.

Red algae do not grow as large as some of the kelps, but in California, a few reach lengths of more than 1 m (3.3 ft). Some of the species, especially epiphytic ones, are quite small, but they are almost all visible to the naked eye. In tropical seas, the Rhodophyta tend to be mostly small forms, but in temperate waters such as ours, the average size is from 5 to 30 cm (2–12 in).

One of the most interesting features of these plants is their color. Although they contain green chlorophyll, that pigment is generally masked by other, so-called accessory, pigments, especially the red phycobilins. In brightly lighted intertidal habitats, the "mix" of pigments is so varied that dark purplish, olive, brownish, or blackish colors are observed and the beginner in phycology may assume some of the dark-pigmented or greenish forms to be brown or green algae. In well-shaded places or in deeper water, the red pigments predominate and the plants are almost invariably purple, pink, or red. This reflected red color results from the algal pigments absorbing green and blue light. This absorption aids growth in the dim and limited light of deep water, for water absorbs red, orange, and yellow light very quickly and allows only the green and blue to penetrate to depth. The red pigment is able to absorb this deeply penetrating light and permits some red seaweeds to live at depths greater than 90 m (300 ft). Since little or no red light is available at depths below 3 m (10 ft) and greens and blues are absorbed, red algae growing in the subtidal often appear black unless illuminated with artificial light.

This mix of pigments also produces some interesting colors as these algae decay. Drift plants commonly turn an almost fluorescent orange as the chlorophyll degrades, leaving behind only the phycobilins. Since these latter pigments are water soluble but chlorophyll is not, you can produce the same fluorescent orange color in water by grinding up a leafy red alga in fresh water and filtering out the pieces of tissue.

Most red algae have a complex life history that includes not one, but three different plants. Thus, unlike the larger brown algae that usually have only one conspicuous plant in their life histories, the red algae have three, the sporophyte, a male, and a female plant, and all of these are generally of similar size and appearance (p. 22, fig. 10). Moreover, the female plant bears an additional spore-producing phase. This life history provides for many complexities in classification and a critical identification of a given species may require the presence of the male, the female, and the two spore-producing phases. In some species, one can recognize these different phases easily, but in most, careful microscopic examination is necessary. Female red algae are most readily distinguished, since masses of carpospores that are produced after fertilization are usually visible as dark spots or lumps in or on the thallus. These spores with the female tissue around them are generally referred to as *cystocarps* (fig. 10).

Sporophytic plants are most commonly encountered in the majority of species and usually bear tetraspores. These are often produced on specialized branches or bladelets or in visible patches (sori) on the thallus. Male plants are generally the least abundant and their minute spermatia can be seen only under the microscope. These nonmotile male gametes require water motion for transport to the female.

In addition to the beauty of their varied forms and colors, the red algae are of considerable economic value. In many parts of the world, various species are used as food, fertilizer, and as a source of industrial gums, including agar, the jelling agent widely employed as a substratum for culturing bacteria. In addition to these current uses, the recent discoveries of antiviral, insecticidal, and herbicidal activity in various red algal extracts suggest that these sea "weeds" will become of even greater importance to man in the future.

Bangia (after the Danish botanist, N. H. Bang)

Bangia fusco-purpurea most commonly occurs as dark purple to black hairlike streaks or mats growing on rocks or wood in the high intertidal zone (pl. 7b). An examination of a bit of this material with a hand lens will show the mat to be composed of many long, cylindrical plants that are filamentous near the base but many cells thick at the tips (fig. 27b). Each plant is attached to the substratum by tiny rhizoids growing from the basal cells. *Bangia* mats are especially common in fall and winter. Grazing by limpets and periwinkles can be particularly intense in the high intertidal zone (pl. 7b) and the decreased activity of these snails in winter, as well as reduced desiccation, may be responsible for *Bangia*'s seasonal occurrence.

Distribution: *B. fusco-purpurea*, entire coast.

Rhodochorton (rose grass)

A close examination of shaded high intertidal rocks and cave interiors will often reveal the presence of dark red masses that feel like fine felt. These masses are the densely aggregated filaments of *Rhodochorton purpureum* (fig. 27a). The plant is actually composed of two types of filaments, prostrate ones forming a basal mat

and upright ones forming a "felt." The dark red color
is probably an adaptation to low light conditions and
this, combined with the plant's ability to withstand long
periods out of water, allows it to exploit a habitat un-
favorable to most other intertidal plants.

In addition to *Bangia* (p. 97) and *Rhodochorton*,
a variety of other small, filamentous reds such as
Goniotrichum, *Erythrotrichia*, and *Acrochaetium* are
very common throughout the intertidal and subtidal
zones, living on rocks or other organisms. Most of these
do not form dense aggregations, however, so are not
apparent unless surfaces are examined with a hand lens
or dissecting microscope.

Distribution: *R. purpureum*, Oregon border to
San Diego.

Smithora (after the American phycologist, G. M. Smith)

Although our abundant intertidal surf grass is usually
emerald green, from time to time it takes on a mottled
purplish color, which is the result of being covered with
the small epiphytic alga, *Smithora naiadum* (fig. 27*c*).
Smithora is an epiphyte on *Phyllospadix* and *Zostera*
(p. 157) on whose leaves it forms abundant purplish,
membranous blades that arise from small, fleshy cushions. It was formerly thought to represent a small species of *Porphyra* (p. 98), but when the life history was
worked out, it proved to have a distinctive and complex
reproductive cycle. It is seasonal and periodic and
sometimes only the tiny reddish cushions will be observed dotting the leaves.

Smithora was named to honor Gilbert Morgan
Smith, professor of botany at Stanford University from
1925 to 1950. Professor Smith wrote the first widely
available comprehensive marine algal flora for central
California and his publications on marine and fresh
water algae are still extensively used.

Distribution: *S. naiadum*, entire coast.

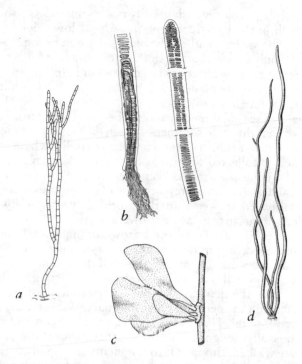

Fig. 27. *a, Rhodochorton purpureum*, branch (1 mm, 0.04 in); *b, Bangia fusco-purpurea*, lower filamentous and upper multicellular portion of a single plant (0.03 mm, 0.001 in); *c, Smithora naiadum* on surf grass (2 cm, 0.8 in); *d, Nemalion helminthoides* (15 cm, 6 in)

Porphyra (purple) Laver, Nori

Porphyra is a red algal genus with many species. These primarily intertidal plants have membranous blades only one or two cells thick. There are several species along the California coast, but only the three most common ones are described here.

P. perforata is a ruffled, stipeless blade, greenish-purple in color, which usually occurs on rocks or shells high in the intertidal zone (pl. 6*b*). If low tides occur during warm weather, the blades dry and become brittle, only to rehydrate with the incoming tide (pl. 6*c*). *P. lanceolata* is another, larger rock-inhabiting species with tapering blades, and *P. nereocystis* lives epiphytically on old *Nereocystis* (p. 70) stipes and floats. These are all seasonal species and are best developed from late winter through summer. *Porphyra* has a gelatinous, somewhat rubbery texture when partially dehydrated, and is collected for use as a foodstuff known as nori. The Japanese nori has been used as food for more than 1,000 years and is presently one of the most extensively cultivated seaweeds of the world; it provides for an enormous marine aquacultural industry.

In Japan, *Porphyra* is grown in shallow bays and, more recently, in deep water. In bays, millions of bamboo poles are driven into the mud during autumn low tides and miles of coarse-mesh nets are strung on them so that the nets are exposed at low tide. Nori spores settle on the nets and grow into harvestable blades in a few months. Harvesting is done both by hand and with specially designed machines. Once ashore, the plants are washed, chopped into fragments and dried into thin sheets slightly smaller than a piece of typing paper, and packaged for market. The discovery that the spores are produced by a filamentous, shell-boring phase, formerly thought to be the separate genus *Conchocelis*, revolutionized nori culture. Fishermen can now artificially "seed" their nets with "*Conchocelis*" spores produced in marine greenhouses, thus assuring a good crop every year. The English botanist, Dr. Kathleen Drew Baker, who worked out this unusual algal life history, is commemorated by a monument on Hiroshima Bay.

In deep water, methods have been developed for growing the plants on huge floating rafts, greatly increasing the size of the culture grounds. These and other discoveries and improvements in technique have increased the annual production to more than 4 billion dried sheets.

If you are interested in trying nori, it can be purchased in specialty markets or you can collect your own. Fresh material should be washed in fresh water, chopped, and dried. It can then be added to salads, stews, soups, and other dishes. One of the most common foods prepared with nori is sushi, a sort of seaweed "taco" of flavored rice, vegetables, and strips of fish wrapped in nori. Several Japanese and Pacific coast Indian recipes can be found in some of the books listed in the bibliography.

Distribution: *P. perforata*, entire coast; *P. lanceolata* and *P. nereocystis*, Oregon border to Point Conception.

Nemalion (thread)

As the specific name suggests, *Nemalion helminthoides* is a very wormlike alga (fig. 27d). The soft, elastic, golden brown to brownish-red plants are commonly found growing in groups on high intertidal rocks occupied by a variety of grazing animals but few other macroscopic plants. The related *Cumagloia andersonii* (pl. 6d) is found in similar habitats and is also soft and slippery but tougher and less elastic. In addition, *C. andersonii* is covered with small branchlets not found on *Nemalion*.

Distribution: *N. helminthoides* and *C. andersonii*, entire coast.

Gelidium (forming a gel when cold) Agarweed Pterocladia (feather branch)

Gelidium and *Pterocladia* are common but extremely variable genera and the range of size among several of their common members plus their resemblance to each other make distinctions difficult. Both genera contain tough, cartilaginous plants that are usually distichously

Fig. 28. *a, Gelidium coulteri*, upper portion of branch (2 cm, 0.8 in); *b, Gelidium purpurascens* (8 cm, 3.1 in); *c, Pterocladia capillacea*, upper portion of branch (3 cm, 1.2 in)

branched. Branches are slender and compressed to flattened. The branches of *Pterocladia* are generally flat and constricted at the junction with the main axis. *Gelidium* branches are more oval, usually lack constrictions, and often bend upward near the junction with the main axis, like an arm bent at the elbow.

An abundant mid intertidal zone species of *Gelidium* is *G. coulteri*, a small, tufted, densely branched plant between 3 and 8 cm (1–3 in) tall (fig. 28*a*). A smaller, somewhat similar species, *G. pusillum*, occurs throughout the intertidal and into the subtidal zone. *G. purpurascens* (fig. 28*b*) can be found in the low intertidal and shallow subtidal zone. It is 8 to 20 cm (3–8 in) tall and more sparsely branched than *G. coulteri*. The largest and probably the most easily identified species is *G. robustum* (pl. 9c). The fronds can be up to 1 m (over 3 ft) long and are compressed and highly branched near the apex, but cylindrical and generally unbranched below. This alga occurs from the low intertidal zone into the subtidal zone in areas of moderate to high water motion. It is often found with the large, wiry, irregularly branched *G. nudifrons* in southern California. *Pterocladia capillacea* (fig. 28c) extends throughout the intertidal and into the subtidal zone in southern California and can be locally very abundant. The flat axes and often regular distichous branching help distinguish it from species of *Gelidium*.

Both genera are sources of high-grade agar used, among other things, as a culture medium for microorganisms and for making soft casts such as dental impressions. *Gelidium robustum* produces very high quality agar and was extensively harvested by divers in southern California during World War II, when agar supplies from Japan were cut off. Agar was such a vital commodity during the war that *Gelidium* divers were exempt from the draft. With recent increases in the price of agar, harvesting has again begun in southern California with about 100 dry tons collected each year. Our *Gelidium* beds are relatively sparse and slow growing, however, so harvesting is still an economically marginal enterprise. American Agar Company of San Diego, the largest agar producer in the United States, acquires most of its raw material from foreign sources.

Distribution: *G. coulteri*, *pusillum*, *purpurascens*, and *robustum*, entire coast; *G. nudifrons* and *P. capillacea*, Santa Barbara to Mexican border.

Constantinea (standing firm)

Perhaps the most easily recognized of all the red algae is *Constantinea simplex*, with its stiff stipe and centrally

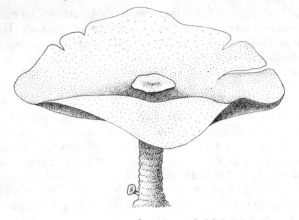

Fig. 29. *Constantinea simplex* (6 cm, 2.4 in).

attached (peltate) leathery blade (fig. 29). The stipe projects upward through the blade and new blades are initiated just above the old blade as the old blade is lost below. This annual cycle of blade loss and initiation is controlled by photoperiod; new blades are initiated only during short winter days. When old blades are lost, they leave scars on the stipe similar in appearance to the bud-scale scars on branches of woody plants. Counting these scars reveals that some *C. subulifera* plants are more than fifteen years old.

   Maripelta rotata and *Sciadophycus stellatus*, the two other local red algae with peltate blades, also deserve mention although they are uncommon. These plants are among the deepest growing of all our marine algae, occurring at depths well below 30 m (100 ft), where only dim, bluish-green light is available for photosynthesis. In the laboratory, both *M. rotata*, which resembles a small *Constantinea*, and the star-shaped *S. stellatus* are positively phototropic, orienting their blades toward the strongest source of light. *M. rotata* can be grown in green light and grows at light intensities nearly 1,000 times less than found at the sea surface.

Distribution: *C. simplex*, Oregon border to Point Conception; *M. rotata*, Monterey to Mexican border; *S. stellatus*, Santa Barbara (Channel Islands) to Mexican border.

Peyssonnelia (after S. A. Peyssonel)

A variety of crustlike, noncalcareous red algae are found throughout the intertidal and subtidal zones, with colors ranging from brownish to rose to dark purplish-red. Most of these crusts are difficult to collect intact and difficult to identify. One of the most distinctive is *Peyssonnelia meridionalis*, which lives primarily on the shells of the brown turban snail (*Tegula brunnea*), giving them a brownish-red color and a rubbery texture (fig. 30c). Sometimes all but the opening of the shell is overgrown. That this covering is the alga and not the outer layer of the shell is evident if the shell is scraped with a fingernail. The alga comes off but the orangish-tan shell will not.

Hildenbrandia spp. and "Petrocelis spp." form rose-red to purplish-black coverings on rocks in the intertidal zone. The dark, slippery crusts of "P. middendorffii" are especially common in the vicinity of Gigartina papillata and G. agardhii (p. 128) and culture studies have shown that the crusts are the sporophytic stage in the life history of Gigartina papillata and probably G. agardhii. Like "Halicystis" (p. 56), "Petrocelis" is placed in quotes because it is no longer valid. Tetrasporangia can be seen in a reddish layer just beneath the surface by crushing out a small bit of the crust on a slide and observing it microscopically (fig. 30d). The presence of tetrasporangia will also assure that you have "Petrocelis" and not the crustose brown alga *Ralfsia* (p. 00).

"Petrocelis" crusts grow very slowly and growth rates on plants in Washington indicate that a 200 cm^2 (31 in^2) crust may be more than ninety years old.

Distribution: *P. meridionalis*, Oregon border to

Point Conception; *Hildenbrandia* spp. and "*P. midden-dorffii*," entire coast.

Pseudolithophyllum (false stone leaf)
Crustose or Encrusting Corallines

Among the most common of all the marine algae are the rock- and shell-inhabiting crustose coralline algae belonging to the genera *Pseudolithophyllum*, *Lithophyllum*, and *Lithothamnium*. Representatives of one or more of these genera occur from the Arctic and Antarctic through the tropics, from Greenland fjords to Caribbean reefs, from intertidal pools to the lowermost limits of light at depths to 150 meters (500 ft). The purple or pink plants may grow as thin, calcareous films only a few cells thick or as knobby, stony masses several centimeters thick, widely spreading on rocks. Sometimes, the crustlike holdfasts of articulated corallines like *Corallina* (p. 111) may be mistaken for these truly crustose genera. Reproductive structures, however, are never found on holdfasts.

   Pseudolithophyllum neofarlowii (pl. 6e) is one of the most conspicuous and abundant intertidal crustose corallines along the California coast. Especially common in mid and high intertidal zone habitats, it occurs as whitish to purple masses up to 1 mm (0.04 in) thick, covering rocks, barnacles, and other available hard substrata. The surface is very irregular and covered with small protuberances.

   Lithothamnium californicum and *Lithophyllum imitans* are two other common species found throughout the intertidal and subtidal zones. The former occurs as smooth, expanded, light purple crusts about 1 mm (0.04 in) thick (fig. 30b). The more nodular *L. imitans* has the appearance of a dark pink ceramic glaze (fig. 30e). These and other encrusting corallines are usually covered with small, moundlike reproductive conceptacles visible with the unaided eye or with a hand lens. Unfortunately, most genera and species can be satisfac-

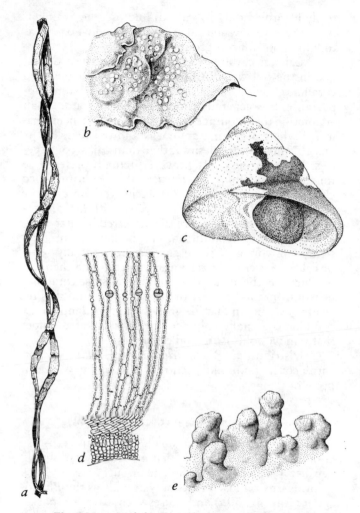

Fig. 30. *a* , *Melobesia mediocris* on surf grass (15 cm, 6 in); *b*, *Lithothamnium californicum* (2 cm, 0.18 in); *c*, *Peyssonellia meridionalis* (dark area) on a turban snail (3 cm, 1.2 in); *d*, "*Petrocelis*" sp., cross section showing tetrasporangia (0.5 mm, 0.02 in); *e*, *Litophyllum imitans* (1.2 cm, 0.5 in)

torily identified only by careful microscopic examination. There are few specialists in the world who profess to know this difficult group.

Calcium carbonate or limestone is actively deposited in the cell walls of both crustose and articulated corallines, giving them their characteristic stony appearance. It is generally thought that this feature is an adaptation to discourage grazing and, indeed, in experiments where a grazing animal is allowed to select its "meal," most prefer fleshy rather than calcareous species. A few marine snails, however, actually prefer these tough plants. *Tonicella lineata*, the beautiful lined chiton, lives and feeds almost exclusively on crustose corallines (pl. 7*a*) and cobbles covered with *Lithothamnium* and *Lithophyllum* are the preferred habitat of juvenile abalone. In addition, the planktonic or free-swimming young of these grazers preferentially settle on calcareous crusts, a response triggered by a chemical produced by the algae. Whether our crustose corallines benefit from this association is unknown, but it has recently been shown that the growth of one Atlantic coast coralline is actually enhanced by an associated limpet that removes old tissue and epiphytes.

Distribution: *P. neofarlowii* and *L. californicum*, entire coast; *L. imitans*, central California to Mexican border.

Melobesia (one of the mythical sea nymphs)

Melobesia mediocris is a small, calcareous, crustose coralline alga that can be identified with ease simply because it is an obligate epiphyte on surf grass (p. 161). It forms tiny pinkish crusts only as wide as the *Phyllospadix* leaf (fig. 30*a*) and causes such leaves to sparkle with these glistening light spots in sunlit water. Another common species, *M. marginata*, lives on other red algae including *Laurencia* (p. 108) and *Gymnogongrus* (p. 125).

Distribution: *Melobesia* spp., entire coast.

Mesophyllum (middle leaf)

Mesophyllum conchatum is readily recognized, for it grows epiphytically on *Calliarthron* (p. 113) and other articulated corallines as circular or semicircular plates up to 2.5 cm (1 in) in diameter attached by the middle of the lower face (fig. 31a). A second species, *M. lamellatum*, is also an epiphyte on articulated corallines, but forms overlapping lobes rather than single plates. Other epiphytic encrusting coralline genera can be found in California but, as is true of crustose corallines in general, their identification requires careful microscopic examination.

Distribution: *M. conchatum*, Oregon border to Point Conception; *M. lamellatum*, entire coast.

Lithothrix (stone hair)

This genus contains the single species *Lithothrix aspergillum*, which is common in sandy areas from the shallow subtidal zone up into intertidal pools. It is often densely branched, tufted, and resembles a coarse shock of purple bristles. The slender branches, less than a millimeter (0.04 in) in diameter, are composed of a great many minute, stony segments about as long as wide, and these bear lateral conceptacles (fig. 31b). Although originally described from Vancouver Island in 1867, this plant shows its most abundant growth along the warm shores of southern California.

Male, female, and sporophytic plants of the various articulated corallines are so similar that one must usually look at the contents of the conceptacles for tetraspores, carpospores, or spermatia to distinguish the three different plants in the life history.

Distribution: *L. aspergillum*, entire coast.

Jania (one of the mythical sea nymphs)

These tropical articulated corallines reach into California only in warm water areas. *Jania adhaerens* (fig. 31c)

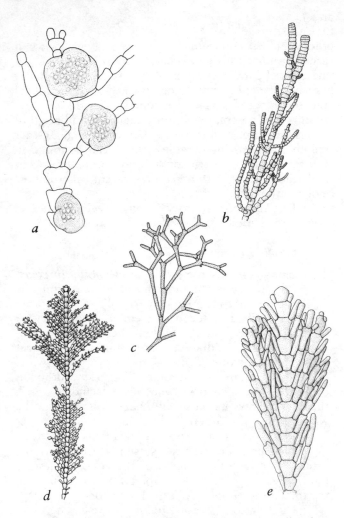

Fig. 31. *a, Mesophyllum conchatum* on *Calliarthron* (8 cm, 3.1 in); *b, Lithothrix aspergillum,* portion of a plant (5 cm, 2 in); *c, Jania adhaerens* (2 cm, 0.8 in); *d, Corallina officinalis* var. *chilensis,* portion of branch (6 cm, 2.4 in); *e,* C. *vancouveriensis,* portion of branch (2 cm, 0.8 in)

and *J. tenella* occur in mats or clumps generally less than a centimeter (0.4 in) thick on rocky substrata or on other algae. *J. crassa*, a larger plant found in South Africa, New Zealand, and California, forms clumps in tide pools. The slender, cylindrical, dichotomously branched fronds are gray-pink and 3 to 5 cm (1–2 in) tall. *Amphiroa zonata* often occurs in the same pools and its overall form is similar to *Jania* spp. The intergenicula, however, are around 1 mm (0.04 in) in diameter rather than 0.1 to 0.4 mm (0.004–0.02 in), typical of *Jania* spp.

Distribution: *J. adhaerens*, San Diego to Mexican border; *J. crassa*, Santa Barbara to Mexican border; *J. tenella*, Catalina Island and San Diego to Mexican border; *A. zonata*, Catalina Island and Oceanside to Mexican border.

Corallina (small coral)

All the calcareous red algae are commonly referred to as corallines, but only one of the many genera actually bears the name *Corallina*. It is our most widespread and prevalent articulated calcareous alga (pl. 2c), readily recognized by its pinnate branching (fig. 5), which is usually more or less distichous, and by reproductive conceptacles that are terminal on the segments. Two other genera (*Haliptylon* and *Serraticardia*) also have some of these characteristics, but both are relatively uncommon. *Corallina*, like all articulated corallines, is provided flexibility against wave shock by the minute uncalcified pads (genicula) between the stony segments.

One of the most common of all our intertidal algae is *Corallina vancouveriensis*, which forms dense tufts and mats through the middle and lower intertidal zones (pl. 7d, fig. 31e). It occurs in a variety of forms that tolerate a wide range of environmental conditions from the Aleutian to the Galapagos Islands. A larger species of tide pools and the subtidal zone is *C. officinalis* var. *chilensis* (fig. 31d). This is the medicinal coralline of the ancients, named by Linnaeus for its favored use as a vermifuge during many centuries before 1775.

The hard bodies of these red algae and the general resemblance, particularly of the crustose corallines, to some limestone formations and to calcareous coral animals caused confusion among early naturalists. Although Linnaeus described the genus *Corallina* in 1758, he thought the plants were corallike animal forms. The famous evolutionary biologist, Lamarck, also thought the corallines were corals but, because they lacked the large openings typical of coral animals, called them nullipores. It was not until 1837 that Phillipi finally recognized that all the corallines were plants.

Distribution: *C. vancouveriensis* and *C. officinalis* var. *chilensis*, entire coast.

Bossiella (after the Dutch phycologist, Madame Anna Weber vanBosse)

Bossiella is distinguished from other articulated corallines by its calcified segments, which have a distinctive "wing nut" shape with the conceptacles being borne on the faces of these flat segments. The dichotomously branched, conspicuously winged *B. orbigniana* ssp. *dichotoma* is probably the most prevalent intertidal species (fig. 32*a*). It is particularly abundant in exposed habitats where, along with other corallines, it forms a pinkish-purple band in the low intertidal zone. *B. californica* ssp. *californica*, more irregularly branched with blunt wings (fig. 32*b*), is also common in the same habitat. *B. californica* ssp. *schmittii*, a subtidal species found in kelp forests, produces fronds that grow parallel with, and close to, the bottom. The clean upper surface is dark purple and bears the conceptacles, while the lower surface is generally pink, lacks conceptacles and serves as a substratum for a variety of sessile animals.

The pink or purple color of all live articulated corallines is quickly lost if they are left to dry in the sun, and one commonly finds these jointed coralline "skel-

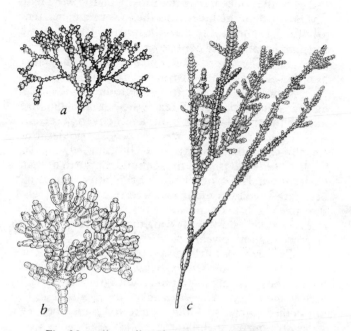

Fig. 32. *a*, *Bossiella orbigniana* ssp. *dichotoma* (10 cm, 4 in); *b*, *B. californica* ssp. *californica* (10 cm, 4 in); *c*, *Calliarthron cheilosporioides* (15 cm, 6 in)

etons" made chalky white by sun bleaching in old beach drift.

Distribution: *B. orbigniana* ssp. *dichotoma* and *B. californica* ssp. *schmittii*, entire coast; *B. californica* ssp. *californica*, Bodega Bay to Mexican border.

Calliarthron (beautifully jointed)

The two species of *Calliarthron* are the largest of all the articulated corallines in California, reaching lengths of more than 20 cm (8 in). *C. cheilosporioides* (fig. 32*c*, pl. 10*e*) resembles a large, coarse, pinnate *Bossiella* (p.

112), but the conceptacles are mostly on the wing margins rather than the faces, and the lower segments tend to be more rounded. *C. tuberculosum* (pl. 10*f*) is coarse, stiff, and usually more densely branched than *C. cheilosporioides*. The upper intergenicula tend to be flat or oval and generally lack distinct wings.

Although these plants can be found on the shore during "minus" tides, they reach their greatest development in the subtidal zone, completely covering the bottom in some areas. *Calliarthron cheilosporioides* is especially common in southern California kelp forests, while *C. tuberculosum* is most abundant north of Point Conception. The growth rate of the latter species has been measured by comparing successive photographs of the same plants taken for more than a year. The average rate is relatively slow, indicating that a 20-cm plant may be more than nine years old.

Calliarthron harbors a variety of epiphytes, including the coralline *Mesophyllum* (p. 109). One small, filamentous red algal epiphyte with the large name *Ptilothamnionopsis lejolisea* takes up residence in the uncalcified joints, forming furlike red bands between successive purple segments.

Distribution: *C. cheilosporioides*, *C. tuberculosum* and *P. lejolisea*, entire coast.

Endocladia (branching within)

The dense, bushy, brownish-red clumps of *Endocladia muricata* are a conspicuous feature of almost all high intertidal rocky shores from Alaska to California (pl. 3*a*). The stiff, cylindrical, spine-covered plants (fig. 33) often grow with *Gigartina papillata* (p. 129) and both can be observed during all but the highest tides. The bushy habit of *Endocladia* aids in trapping moisture and the relatively cool, moist interiors of the clumps serve as a low tide refuge for more than ninety species of animals in central California, including snails, crustaceans, worms, and even the larvae of some beach flies.

Fig. 33. *Endocladia muricata*, vegetative branches on left (3 cm, 1.2 in) and female branches on right (1.5 cm, 0.6 in)

The clumps are also a preferred settling site of mussels. The mussels may eventually crowd out the alga, but it can reestablish by settling on top of the mussel shells.

Distribution: *E. muricata*, entire coast.

Halymenia (sea membrane)

A variety of generally unbranched, flat, foliose red algae grow in our low intertidal and subtidal waters but, unfortunately, the similarities in external form make identification difficult without a careful examination of their internal structure. *Halymenia* falls into this category, although the usually rose-red and soft or slippery blades aid in distinguishing it from other plants of similar form. The most common species, *H. californica* (fig.

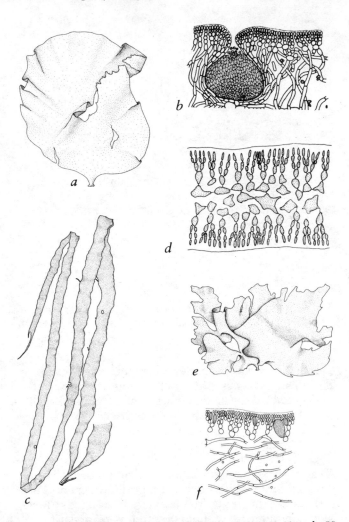

Fig. 34. *a, Halymenia californica* (40 cm, 16 in); *b, H. californica*, upper half of cross section through cystocarp (0.3 mm, 0.01 in); *c, Grateloupia doryphora* (40 cm, 16 in); *d, G. doryphora*, cross section (0.1 mm, 0.004 in); *e, Schizymenia pacifica* (30 cm, 12 in); *f, S. pacifica*, upper half of cross section showing large gland cells (0.2 mm, 0.008 in)

34*a*, 34*b*), occurs as soft, dark rose-red blades up to 70 cm (27 in) tall. It is a strictly subtidal plant, occurring to depths of 30 m (98 ft).

Distribution: *H. californica*, entire coast.

Schizymenia (split membrane)

Schizymenia pacifica is a very slimy, brownish-red alga of the subtidal zone and lowermost, wave-exposed rocks. One can learn to recognize *Schizymenia* by touch. The thick, smooth, membranous blades are usually broad and 30 to 60 cm (1–2 ft) long, but much split or lacerated (fig. 34*e*, 34*f*). They arise from a very short stipe and small holdfast. The plants can be distinguished anatomically from other foliose red algae by the presence of large, oval gland cells (fig. 34*f*) seen when blade cross sections are viewed under the microscope.

Distribution: *S. pacifica*, entire coast.

Grateloupia (after the French naturalist, Grateloup)

The low intertidal zone in both sheltered and exposed locations is often occupied by soft, wine-red to purple blades of *Grateloupia doryphora*. Blades of this highly variable species are generally long and narrow (fig. 34*c*, 34*d*), but some are linearly divided above a flat stipe. Blade margins on both of these types may be smooth or covered with proliferations ranging in size from small (1 mm or 0.04 in wide) and spinelike to large (more than 2 cm or 0.8 in wide) and bladelike. This variation can make identification difficult and some forms require microscopic examination to separate them from *Schizymenia* (p. 117) and *Prionitis* (p. 117).

Distribution: *G. doryphora*, entire coast.

Prionitis (serrated)

This genus includes several species that characteristically occur at low tide levels and below, or in deep tide

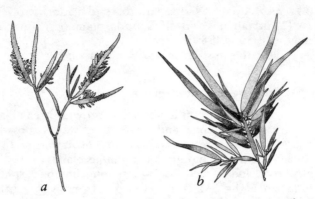

Fig. 35. *a*, *Prionitis lanceolata* (25 cm, 10 in); *b*, *P. lyallii* (30 cm, 12 in)

pools. They are coarse plants with narrow, compressed branches, often more than 30 cm (1 ft) long, being deep, dull purplish-red in color and having a cartilaginous texture. The common species, *P. lanceolata*, is highly variable; figure 35*a* shows a plant typical of the mid to low intertidal zone. Subtidal forms usually lack proliferations, while lightly colored, irregularly branched plants with numerous proliferations are found in high intertidal pools. *P. lyallii*, a resident of sandy habitats, has numerous distichously arranged, lancelike blades (fig. 35*b*). A characteristic feature of most species is the presence of irregular series of very short, determinate, distichous branchlets along the axes and main branches. Plants also often have a fresh, chlorinelike smell.

Distribution: *P. lanceolata* and *P. lyallii*, entire coast.

Erythrophyllum (red leaf)

When fully developed in midsummer, the dark red blades of *Erythrophyllum delesserioides* closely resemble delicate, elongate, dark red leaves (fig. 36). This

Plate 1

▲ B Various lichens on volcanic rock above the intertidal zone

◄ A The green alga, *Trentepohlia*, on cypress

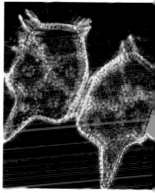

▲ C *Asterionella*, a diatom which forms rings of cells

▲ D *Dinophysis*, a marine dinoflagellate. Two cells stuck together after division

◄ E Floating *Enteromorpha* spp. mat in Elkhorn Slough

Plate 2

▲ **A** Drifting *Gracilaria sjoestedtii* and *Ulva* spp. in San Francisco Bay

▲ **B** Algal zonation near Oakland. A black zone of blue-green algae with two green zones of *Ulva* spp.

▲ **C** Zonation of dark *Gigartina* and pink *Corallina* at La Jolla

▲ **D** *Lessoniopsis littoralis* and *Laminaria dentigera* on the exposed coast near Carmel Bay

▲ **E** Intertidal *Ulva* (leafy green), *Colpomenia* (brown) and *Cladophora* (green clumps)

▶

D A *Macrocystis pyrifera* forest at Santa Cruz Island

Plate 3

▲ A Intertidal zonation at Monterey. From high to low: *Prasiola,* goose barnacles, *Endocladia* (brown clumps), *Iridaea* (green), *Gigartina* (red) and surf grass (green)

▲ B Surf grass and *Laminaria denti gera* at low tide

◄ C Subtidal *Eisenia arborea*

▼ E *Egregia* at low tide

Plate 4

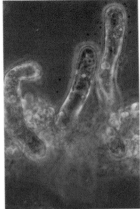

▲ B Giant kelp gameto-
phytes 400x

▲ A The Sea Palm

▼ C Giant kelp sporophytes near the surface

Plate 5

▲ **B** *Fucus* (large fronds) and *Pelvetia*

◄ **A** Bull Kelp. Note the relatively small holdfast

▲ **C** Elk Kelp in the subtidal zone at Catalina Island

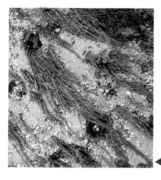

◄ **D** *Scytosiphon lomentaria* and barnacles

Plate 6

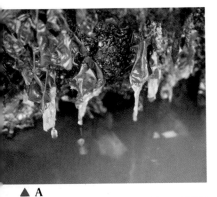

▲ A

▲ **B** *Porphyra* just after exposure to air

A *Ulva* with yellow areas indicating release of reproductive cells

◀ **C** *Porphyra* dried after exposure to air on a hot day.

▲ **D** *Cumagloia andersonii* and limpets in the splash zone near Half Moon Bay

◀ **E** *Pseudolithophyllum neofarlowii* (light purple) and a limpet on intertidal rock

Plate 7

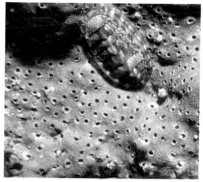

▲ **A** The chiton, *Tonicella lineata*, on a crustose coralline. The holes in the coralline are made by worms.

▲ **B** *Bangia* mat grazed by limpets and periwinkles

▶

E *Gigartina papillata*

▼ **D** A clump of *Corallina vancouverienses*

C The Garibaldi guarding his nest of red algal turf. The yellow area is a mass of eggs.

▼

Plate 8.

▲ A

▲ B *Plocamium cartilagineum*

A *Microcladia coulteri*
on *Gigartina*

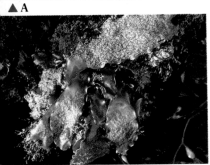

◄ C *Gigartina corymbifera*

◄
D *Gigartina harveyana*

Plate 9

◀ A A clump of *Gigartina canaliculata*

▲ B *Gymnogongrus platyphyllus*

▼ C *Gelidium robustum*

D A branch of *Gastroclonium coulteri*
▼ showing hollow branchlets

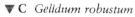

Plate 108

▲ B *Botryocladia pseudodichotoma*

▲ A *Rhodymenia pacifica*

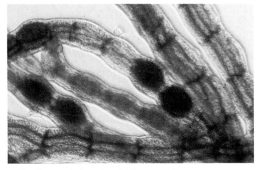

▲ D *Pterosiphonia* 100x. Dark areas are tetrasporangia.

▲ C Antithamnion 100X

▼ E *Calliarthron cheilosporioides*

▼ F *Calliarthron tuberculosum* and a turban snail

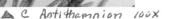

Plate 11

▲ A *Phyllospadix scouleri* with fruiting stem

▲ B *Spartina* and *Salicornia* with patches of orange Salt Marsh Dodder

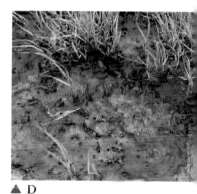

▲ C *Salicornia virginica* with the orange parasite, Salt Marsh Dodder

▲ D

D California Cord Grass with blue-green algal mats on mud

E *Elymus mollis* ▶

Plate 12

A American Dune Grass (above), New Zealand Spinach (below) and sea rocket on foredune in central California

▲ **C** Beach Morning Glory (flowering) and Silky Beach Pea

◄ **B** Marram Grass

▼ **D** Sea Palms in the surf

E Sea rocket growing in driftwood ▼

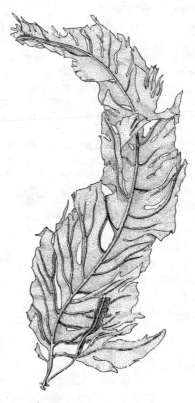

Fig. 36. *Erythrophyllum delesserioides* (10 cm, 16 in)

beautifully veined alga is found along exposed central and northern California shores, where at low tide it commonly hangs down from the undersides of rocky outcrops in the low intertidal zone. In fall and winter, the blades become lacerated and eroded and densely packed reproductive papillae develop along the midrib and lateral veins.

Distribution: *E. delesserioides*, Oregon border to Morro Bay.

Callophyllis (beautiful leaf)

Befitting the name, these plants are among the most brilliantly colored and attractively branched of the red seaweeds. They are usually found in deep waters in which the phycobilin pigments are prominently developed, giving the plants a rose to orange-red color. Intertidally, they are plants of the driftweed and may be best sought after the first winter storms. Most species are flat and thin bladed with a roughly fan-shaped outline. *C. flabellulata* is a common one recognized by its finely dissected ultimate branches and emergent cystocarps that tend to be arranged around the margins of the blades (fig. 37*c*). *C. violacea* is a narrow species with longer segments (fig. 37*a*). Both of these species are usually less than 10 cm (4 in) tall. *C. pinnata*, a coarser and larger plant, is clearly palmate with relatively broad lower branches.

Callophyllis may be recognized by its cross sectional structure, which is unique among the branched, flat, red algae. Small, angular cells are scattered among the large, parenchymatous cells in the medulla (fig. 37*b*).

Distribution: *C. flabellulata* and *C. violacea*, entire coast; *C. pinnata*, entire coast but rare south of Point Conception.

Neoagardhiella (new; named after the Swedish phycologist, J. G. Agardh)

Neoagardhiella gaudichaudii is a widespread species along Pacific America from Alaska to Chile. We find the plant as a common intertidal zone, tide pool, or subtidal zone species. Cystocarpic plants are easiest to identify. These have slender, long, fleshy, cylindrical branches with few to many lateral branchlets and embedded cystocarps that form bulges on the surface (fig. 38*a*). Sterile plants resemble *Gracilaria* (p. 121), but are generally larger in diameter, stiffer, and more highly branched. The two genera are readily separated when a

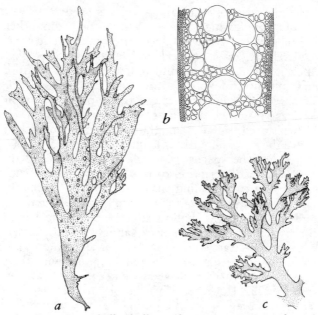

Fig. 37. *a, Callophyllis violacea* (15 cm, 6 in), *b, Callophyllis* sp., cross section (0.3 mm, 0.01 in); *c, C. flabellulata*, portion of plant (4 cm, 1.6 in)

cross section is examined, for *Neoagardhiella* has a central core of fine longitudinal filaments in the medulla (fig. 6a).

J. G. Agardh and his father, C. A. Agardh, were the leading European phycologists in Sweden throughout the nineteenth century. Many of our California seaweeds were described by the younger Agardh from collections made around Santa Barbara and Santa Cruz.

Distribution: *N. gaudichaudii*, entire coast.

Gracilaria (very slender)

This widely distributed and economically important genus occurs in a variety of forms and habitats throughout the world. *Gracilaria* can be both cylindrical and flat. Our most abundant flat species is G. *textorii* var. *cunninghamii*, which can be distinguished from other red algae by its narrow-angled dichotomous branching, relatively large size, and parenchymatous structure (fig. 38c). It is usually deep, dull reddish, fleshy in texture and inhabits tide pools and the subtidal zone. G. *sjoestedtii*, a cylindrical, spaghettilike plant, occurs in bays and estuaries as well as along the open coast. Its brownish-red, slender, lax thallus, sparsely branched above the base, may reach lengths of 2 m (6.5 ft; figs. 10, 38b). The species resembles *Neoagardhiella* (p. 120) and the description of this genus should be consulted before concluding that you do have *Gracilaria*.

Gracilaria sjoestedtii is usually found from the mid intertidal to the shallow subtidal zone, attached to rocks partially buried in coarse sand. In muddy bays, it often grows attached to small bits of clam and oyster shell. In such situations, the plants, along with their substrata, are easily removed by currents and wave action (pl. 2a). Unless cast ashore or carried into deep water, however, they can continue to grow while drifting. In fact, for many algae, spending some time as drift may be an important means of dispersing spores and gametes.

Gracilaria verrucosa is another cylindrical species found most commonly in quiet water. It resembles G. *sjoestedtii*, but is larger and usually has numerous short, spinelike branches in the lower portions. G. *andersonii*, a common, densely branched species found in sandy southern California habitats, has wiry, more oval branches and is generally less than 15 cm (6 in) tall.

Species of *Gracilaria* are agarophytes, second only to *Gelidium* (p. 101) in commercial importance. Because of its high jelling temperature and low viscosity, the agar extracted from most *Gracilaria* is unsuitable for microbiological work. But it is used extensively as a preservative and as a thickening and sizing agent. Various species are used in different parts of the world. G. *verrucosa* was collected commercially from San Diego Bay for a short period during World War II.

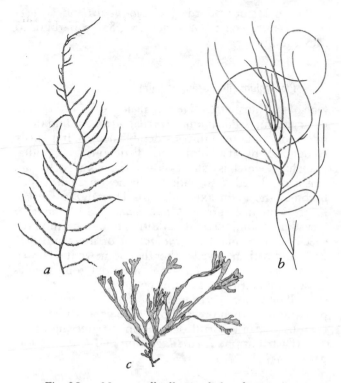

Fig. 38. *a*, *Neoagardhiella gaudichaudii* (35 cm, 14 in), *b*, *Gracilaria sjoestedtii* (30 cm, 12 in); *c*, *G. textorii* var. *cunninghamii* (15 cm, 6 in)

The genus, called ogo in Japan, is also used in a variety of oriental and Hawaiian seaweed dishes. After washing in fresh water, pieces of both *Gracilaria* and *Neoagardhiella* (p. 120) can be added as condiments to salads. Herring lay some of their eggs on *Gracilaria* in San Francisco and Tomales Bays, and a large fishery has recently developed to harvest these "herring eggs on seaweed" for shipment to Japan, where the combination is considered a delicacy.

Distribution: *G. textorii* var. *cunninghamii*, Morro Bay to Mexican border; *G. sjoestedtii* and *G. verrucosa*, entire coast; *G. andersonii*, Santa Barbara to Mexican border.

Plocamium (like braided hair)

One of our most attractive red algae, both in color and in lacy form, is *Plocamium cartilagineum*, abundant in tide pools and down to considerable depths. It is easily recognized by its color and distichous, sympodial branching, which is a succession of zigzags as each main axis is displaced to the side by the next branch, which becomes the main axis (pl. 8*b*). This zigzag pattern helps distinguish it from *Microcladia* (p. 139), another lacy alga of similar size. In addition to the common red species is the smaller *P. violaceum* of more purplish color and with incurved rather than somewhat recurved branchlets. It is a mid to low intertidal zone species most abundant in areas of heavy surf.

   Plocamium cartilagineum makes very fine dry preparations when pressed on white paper. It has long been used as a colorful decoration for greeting cards.

   Distribution: *P. cartilagineum* and *P. violaceum*, entire coast.

Opuntiella (diminutive of Opuntia, a genus of cactus)

Opuntiella californica (fig. 39) is a curiously shaped, rather coarse, cartilaginous alga of deep, dark red color. The broadly rounded, stiff blades that rise from the margin of similar parent blades give the suggestion of the branches of a prickly pear cactus (*Opuntia*). The shape, color, and texture (in combination) make this a markedly distinct plant. It is encountered occasionally in the lowermost intertidal zone, but more often in the subtidal zone.

   Distribution: *O. californica*, entire coast but rare south of Point Conception.

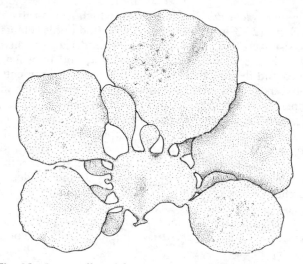

Fig. 39. *Opuntiella californica* (30 cm, 12 in)

Hypnea (mosslike)

Hypnea is a widespread tropical genus that is harvested in many areas for extraction of a substance similar to agar (p. 103). Three species and two varieties occur in southern California, but only one, *H. valentiae* var. *valentiae* (fig. 40d) is common. This brownish-red alga is a summer annual, growing in bushy clumps up to 25 cm (10 in) tall in the low intertidal zone.

Distribution: *H. valentiae* var. *valentiae*, Santa Barbara to Mexican border.

Gymnogongrus (naked swelling)

This is a small genus of flattened plants with regular dichotomous branching and dense structure. Female plants have "naked swellings" called nemathecia. These are cystocarps that form large, wartlike bulges on the surface of the blades. Our most common species is *Gym-*

nogongrus leptophyllus (fig. 40*a*), which forms tufts 5 to 8 cm (2–3 in) tall in sandy, low tide pools. Plants consist of erect axes that are at first cylindrical, attached by a thin, spreading, discoid holdfast, and then flattened and densely dichotomously branched. Segments between dichotomies become progressively shorter in upper parts and tend to be matted and entangled. *G. platyphyllus* is larger than *G. leptophyllus*, with free, erect branches and a very fan-shaped appearance (pl. 9*b*). A third species, *G. linearis*, is similar in form and size to *G. platyphyllus* but with very thick, stiff, brownish-red branches. It is usually found in dense patches on rocks where sandy beaches adjoin rocky areas.

Distribution: *G. leptophyllus* and *G. platyphyllus*, entire coast; *G. linearis*, Oregon border to Point Conception.

Stenogramme (narrow line)

The deep to rose-red *Stenogramme interrupta* ranges from the low intertidal zone to depths of 40 m (130 ft). Its crisp, dichotomous branches resemble those of the more common *Rhodymenia* (p. 126), but the holdfast is discoid and the location of the reproductive structures is distinctive (fig. 40*b*). The cystocarps of female *S. interrupta* appear as a thin, dark, interrupted midrib, while the spore-bearing areas on the sporophyte are distributed in dark blotches scattered over the surface of the blades.

Distribution: *S. interrupta*, entire coast.

Rhodymenia (red membrane)

Color in *Rhodymenia* is true to its name. Always pink or red, these small, dichotomously branched, digitate or fan-shaped blades are commonly found at the lowest tidal levels under overhanging rocks or in clefts and crevices, where the light is subdued and there is limited

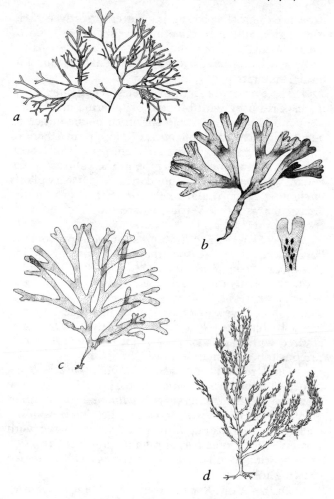

Fig. 40. *a, Gymnogongrus leptophyllus* (8 cm, 3 in); *b, Stenogramme interrupta*, female plant and portion of tetrasporic branch (8 cm, 3 in); *c, Rhodymenia californica* (8 cm, 3 in); *d, Hypnea valentiae* var. *valentiae* (20 cm, 8 in)

exposure to air. The branching pattern, relatively thin, stiff blades, and distinctive internal structure (fig. 6b) aid in distinguishing the genus from *Callophyllis* (p. 120), which can have a similar shape and is found in similar habitats.

There are two common species. *Rhodymenia californica* is bushy with blunted or tapering apices (fig. 40c) and looks like a small specimen of the larger *R. pacifica*, which reaches lengths of more than 10 cm (4 in) and has generally broader blades (pl. 10a). Both range down to moderate depths in the subtidal zone, and they are particularly abundant as understory plants in shallow areas of giant kelp forests and in polluted areas such as Palos Verdes, where they are the most common subtidal alga.

These species produce spreading stolons from the base of the stipe, a feature that aids in distinguishing them from *Stenogramme* (p. 126). The stolons, which allow the plants to spread vegetatively by attaching to other algae, sessile animals, and rock, appear to give some competitive advantage in securing a place to grow. Also, the blades of *Rhodymenia* are themselves often covered with hydroids, bryozoans, and the white calcareous tubes of small worms.

Until recently, the genus *Rhodymenia* also included *R. palmata* var. *mollis*, a broad-bladed alga now placed in the genus *Palmaria*. *P. palmata* is uncommon along our coast, but is abundant in the North Atlantic. Called dulse in Europe, the plant is dried and eaten with various other foods. It is also baked into bread, added as a flavoring to milk, and even used as a substitute for chewing tobacco.

Distribution: *R. californica* and *R. pacifica*, entire coast; *P. palmata* var. *mollis*, Oregon border to Pismo Beach.

Gigartina (grapestone)

The several common members of this genus are often the most prevalent and conspicuous algae in the inter-

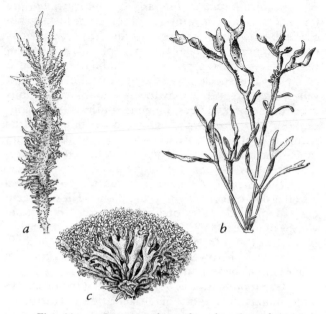

Fig. 41. *a, Gigartina leptorhynchos,* branch (10 cm, 4 in); *b, G. agardhii* (15 cm, 6 in); *c, Rhodoglossum affine,* clumplike mass of branches (7 cm, 2.8 in)

tidal and shallow subtidal zone (pl. 2*c*). Although shape varies from species to species, most can be easily recognized as *Gigartina* by the multitude of papillae covering the surfaces of the blades. *Gigartina papillata* is a dark red to black species, abundant in the high intertidal zone in northern and central California (pl. 7*e*). The apical portion of the thallus is often dichotomously divided, although irregular and undivided plants can also be found. Male plants lack papillae and may resemble *Rhodoglossum* (p. 131). The generally taller, narrower and sparser papillate *G. agardhii* (fig. 41*b*) is found in the same region and habitat. In both species, the sporophyte is probably the crustose "*Petrocelis*" (p. 105).

Gigartina canaliculata and *G. leptorhynchos* are found in the mid to low intertidal zone. The former species is abundantly and narrowly branched with obvious distichous branching near the apex (pl. 9*a*). Papillae develop only when the plant is reproductive. Flattened clumps of this species, which vary in color from light yellow-green to purple, may form an almost continuous cover in the rocky intertidal zone south of Point Conception. *G. leptorhynchos* is densely papillate and irregularly branched, with papillae and small branches often obscuring the cylindrical to bladelike main axes (fig. 41*a*).

The strictly low intertidal to subtidal zone species are all broad-bladed, fairly coarse plants. The oval, generally undivided blades of *Gigartina exasperata* attain widths of more than 20 cm (8 in) and have a pointed or blunt apex (fig. 42*a*). Open coast plants are fairly thick with a "turkish towel" feeling, while plants in calm water tend to be thin and crisp.

G. corymbifera (pl. 8*c*) is the largest of our species and among the largest of all red algae, its broadly rounded blades reaching lengths of more than 1 m (3 ft). These thick, coarse plants with smooth, iridescent bases occur to a depth of 30 m (70 ft). The blades of *G. harveyana* and *G. spinosa* are narrower than the above two species and usually branched. The texture of the former (pl. 8*d*) is similar to open coast *G. exasperata*, while *G. spinosa* is very coarse with large papillae and numerous spinelike to bladelike branchlets (fig. 42*b*). *G. volans* also occurs in the low intertidal zone, but usually in habitats scoured by sand. Its dark, brownish-purple blades are very stiff with few to no papillae. The blade margins may be smooth or covered with round bladelets or elongate papillae, depending on their sex and reproductive condition.

Gigartina is harvested in the North Atlantic and Baja California, Mexico, for its carrageenan (see discussion under *Rhodoglossum*, p. 131). Recent price increases of this valuable seaweed extract have stimulated

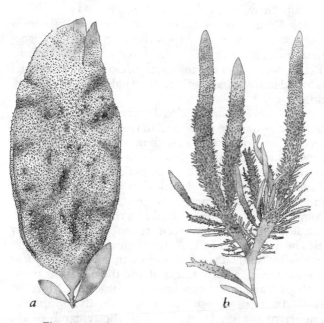

Fig. 42. *a*, *Gigartina exasperata* (30 cm, 12 in); *b*, *G. spinosa* (30 cm, 12 in)

research projects that are currently evaluating the harvest and aquaculture potential of various species of *Gigartina*, *Iridaea* (p. 133), and *Rhodoglossum* (p. 131) in the North Pacific.

Distribution: *G. agardhii*, Oregon border to Pismo Beach; all other species, entire coast but rare south of Los Angeles.

Rhodoglossum (red tongue)

The only common species of this genus, *Rhodoglossum affine*, is, despite the Latin name, hardly to be compared with a red tongue. It is a dull, reddish-brown or purplish, dichotomously branched plant 5 to 10 cm (2–4 in) tall. Growing in dense clumps (fig. 41*c*), it is especially abundant in northern and central California and occupies the same exposed rock habitat as *Gigartina canaliculata* (p. 130). The less common *R. californicum*, which does resemble a thin, elongate red tongue, is found in sheltered low intertidal habitats and the subtidal zone.

Rhodoglossum is closely related to both *Gigartina* and *Iridaea* (p. 133) and to the Atlantic *Chondrus crispus*, world famous as irish moss and the basis of a large seaweed industry. Irish moss was originally harvested in Europe and used in making a kind of milk pudding, but in recent decades has yielded the valuable industrial gum extract, carrageenan. Like agar, this complex carbohydrate is used as a thickening, stabilizing, and jelling agent. Recent research suggests it may also be used as a substitute for the more expensive agar in pharmaceutical work. Large quantities of *Chondrus* are harvested in New England and the maritime provinces of Canada for carrageenan extraction. More recently, the tropical genus *Eucheuma* has also become an important source. Largely due to the research of Dr. Maxwell Doty at the University of Hawaii, successful aquaculture procedures have been developed for this latter genus and *Eucheuma* farms are fast becoming a common feature of shallow Southeast Asian waters.

Distribution: *R. affine* and *R. californicum*, entire coast.

Fauchea (after the French naturalist, Fauche)

Fauchea laciniata (fig. 43*b*) is an example of one of the brightly colored red algae that are so sensitive to air exposure that they seldom are found in intertidal habitats except as drift. In contrast, they are sometimes abundant in deeper waters where the blades take on a beautiful blue-green iridescence. *Fryeella gardneri*, another iridescent subtidal plant, is slightly larger than *F.*

laciniata and has a very thin, hollow central cavity with partitions that appear from the outside as concentric arches. The soft and slippery nature of the thallus and the crown of pointed lobes on the cystocarps of *Fauchea* aid in distinguishing it from *Rhodymenia* (p. 126).

Distribution: *F. laciniata*, Oregon border to San Diego; *F. gardneri*, entire coast.

Iridaea (iridescent)

Along much of the mid to low intertidal rocky shore of northern and central California, the bulk of the red algae is made up of *Iridaea*, for the plants are both large and abundant (pl. 3*a*). They usually consist of several large, smooth, tapering blades up to 1 m (3.3 ft) long, which arise from a common crustose holdfast. Color varies from greenish-olive to deep, rich purple and the blades are rubbery and usually show a marked, oily iridescence when submerged. *I. flaccida* is a greenish species of the mid intertidal zone, while the purple *I. cordata* (fig. 43*a*) occurs in the low intertidal zone, both in areas of moderate wave action. The smaller (around 15 cm or 6 in) and thinner *I. heterocarpa* occurs with *I. flaccida* and up into higher areas. Its blades have irregular margins and female plants are distinguished by very large, bulging cystocarps. One other common low intertidal and subtidal species, *I. lineare*, resembles a small *Laminaria* (p. 84), but has a dark purple stipe and blade.

The blades of *Iridaea* are frequently riddled with holes, giving them a lacy appearance. A close examination of such blades will reveal the presence of the small black snail, *Lacuna*. This tiny molluscan grazer, which lays small (1−2 mm or about 0.06 in in diameter) lime-green egg masses on a variety of intertidal plants, seems to prefer eating reproductive *Iridaea* blades, but can also be found on other algae as well as surf grass (p. 161).

The iridescence for which the genus was named is a result of the multilayered structure of the cuticle that

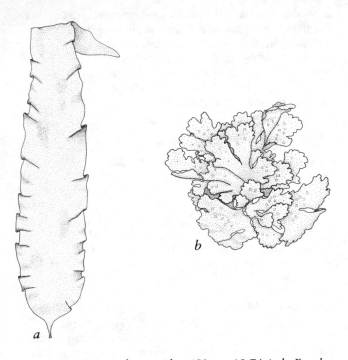

Fig. 43. *a, Iridaea cordata* (50 cm, 19.7 in); *b, Fauchea laciniata* (6 cm, 2.4 in)

covers the outside of the blade. Light reflected from these layers interacts with incoming radiation and the resulting constructive and destructive interference produces the color display. A similar iridescence can be seen when two pieces of glass are pressed together, trapping a thin layer of air between them. The functional significance of this phenomenon in *Iridaea* is unknown; it may be simply an artifact of the structure of the cuticle. The function (or functions) of the cuticle in algae is not well understood. It could inhibit epiphytes, reduce grazing, or help prevent water loss during low tides.

Iridaea flaccida, like many other intertidal plants and animals, shows a rather distinct vertical distribution, rarely occurring above or below the mid intertidal zone. Field experiments indicate it cannot tolerate the more rigorous climate of the high intertidal zone and quickly dies if transplanted into higher areas where *Endocladia* (p. 114) or *Gigartina papillata* (p. 129) grow. But plants will grow and reproduce if transplanted lower. Populations apparently cannot establish because they are eventually outcompeted for space by other algae and sessile animals that normally occupy the lower zone.

In addition to the above, this common genus is also of interest because it produces the chemical, carrageenan (p. 132). *Iridaea* has been commercially harvested in the Pacific Northwest and several investigations are currently under way to devise economically feasible aquaculture techniques for this plant.

Distribution: *I. flaccida* and *I. cordata*, entire coast, but rare south of Santa Barbara; *I. lineare* and *I. heterocarpa*, Oregon border to Santa Barbara.

Gastroclonium (belly branch)

This is a genus identified with ease, since its ultimate branchlets are hollow and provided with diaphragms to form a series of small, slightly swollen chambers (pl. 9 d). In southern California, however, plants of *G. coulteri* tend to bear rather few of these hollow branchlets and to consist largely of cylindrical, fleshy, branched stipe parts. Furthermore, the tender, hollow parts seem to be attractive food for invertebrates and commonly are eaten off to the solid stipe. Accordingly, to make satisfactory identification one must look for well-developed specimens not suffering from grazing. Plants up to 25 cm (10 in) long are found in pools or on the sides of rocks not severely exposed in the low intertidal zone. *Gastroclonium* is more common north of Point Conception and commonly occupies rocks throughout much of the mid to low intertidal zone. *Coeloseira*, a closely related genus that also has chambered branchlets, may also be found. It looks like a diminutive edition of *Gastroclonium* only 5 cm (2 in) high or less.

Distribution: *G. coulteri*, entire coast; *Coeloseira* spp., Monterey to Mexican border.

Halosaccion (salt sack) Sea Sacks, Sea Nipples

One of the most curious of the intertidal red algae is *Halosaccion glandiforme* (fig. 44), which grows as a group of erect, slender sacks about 2 cm (0.8 in) in diameter and up to 25 cm (10 in) long. The plants are usually yellowish to olive-brown if growing in full sunlight, and red to dark purple if living in shade. The sacks are often partially filled with water and, surprisingly, this water sometimes serves as the habitat for tiny crustaceans that can be seen swimming about inside a sea sack if it is held up to the sunlight. When squeezed, these water-filled sacks emit several fine jets of water through apical pores, much to the amusement of those who accidentally discover this property.

Distribution: *H. glandiforme*, Oregon border to Point Conception.

Botryocladia (grape branch) Sea Grapes

After a winter storm, one often finds on the beaches a peculiar seaweed resembling a bunch of red grapes. This is *Botryocladia pseudodichotoma* (pl. 10*b*), which cannot be confused with any of our other algae. It grows primarily in the subtidal zone, often attached to the holdfasts of *Macrocystis* (p. 73). The bladders are not filled with air, but with a clear mucilage. This mucilage is slightly less dense than sea water, allowing the plants to float gently above the substratum.

Distribution: *B. pseudodichotoma*, entire coast.

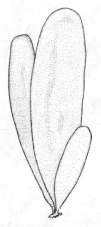

Fig. 44. *Halosaccion glandiforme* (12 cm, 4.7 in)

Antithamnion (opposite shrub)

The genus *Antithamnion* is composed of several small (generally less than 4 cm or 1.6 in tall), very delicate, filamentous species that often occur as red to pinkish, bushy epiphytes on other algae. The beauty of this alga can be fully appreciated only if it is viewed with a micro-scope so that the fine, opposite branches are clearly visible (pl. 10c). *Platythamnion* is of similar size but has four branches on each cell of the main filament rather than two as in *Antithamnion*. It usually grows on rocks or other nonliving substrata. One pair of these branches is always shorter than the other and their arrangement gives the plants a distichous and distinctive banded appearance.

These and related genera as well as species within each genus are often difficult to distinguish without careful microscopic observation and the use of technical references.

Distribution: *Antithamnion* spp. and *Platytham-nion* spp., entire coast.

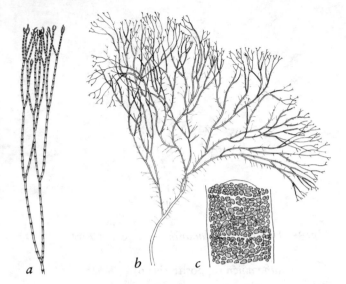

Fig. 45. *a, Centroceras clavulatum* (10 cm, 4 in); *b, Ceramium pacificum* (15 cm, 6 in); *c, C. pacificum*, detail of branch segment (0.3 mm, 0.01 in)

Ceramium (vessel)

Among our delicate forms of marine algae is a genus with many species of small, dichotomous plants of cylindrical form. The main filamentous axes of these algae frequently appear banded to the unaided eye, a pattern that results from the unequal distribution of small cortical cells. Though these cells may cover the entire axis, they are usually concentrated at the nodes or junctions between the axial cells and appear as dark bands, while the areas between are light. Species are often epiphytic, but some form tufts on rocks. They seldom are more than 5 to 8 cm (2–3 in) tall.

Ceramium pacificum is a very common species in California intertidal and subtidal waters. It is densely proliferous and deep red in mass (fig. 45*b*, 45*c*). Another species, *C. codicola*, lives exclusively on *Codium fragile* (p. 54) and has penetrating rhizoids with bulbous tips that fasten it between the host tissue. A common alga that resembles *Ceramium* is *Centroceras clavulatum* (fig. 45a). *C. clavulatum* is found on rocks in the sand-swept low intertidal zone and can be distinguished from *Ceramium* by the whorls of minute spines that project from the cortical cells at each node. The plant tends to fragment at these nodes. *Centroceras* from the Red Sea forms short, missilelike branches that serve as asexual reproductive bodies.

Distribution: *C. pacificum* and *C. codicola*, entire coast; *C. clavulatum*, Santa Cruz to Mexican border.

Microcladia (small branches)

Microcladia coulteri is one of the most common and attractive epiphytes. It is a finely branched, delicate plant of rose-red color that may reach 30 cm (1 ft) in length, being attached to *Gigartina* (p. 128), *Prionitis* (p. 117) or other coarse red algae (pl. 8a). The branching is regularly alternate and distichous. The whole plant has a somewhat pyramidal shape and can resemble *Plocamium* (p. 124). *M. coulteri*, however, usually has a narrower and straighter (percurrent) main axis. It reaches its best development on host plants in subtidal waters and the finest specimens are encountered there and in drift.

The dark, reddish-brown *Microcladia borealis* in habits surf exposed intertidal rocks, commonly growing beneath sea palms (p. 77) in central California. In this species, the branching is pectinate, all branches being produced on one side of the main axis.

Distribution: *M. coulteri*, entire coast; *M. borealis*, Oregon border to Pismo Beach.

Callithamnion (beautiful small branch)

Most species of this genus are small, filamentous plants resembling *Antithamnion* (p. 137). *Callithamnion pike-*

anum is the exception, however, with a thallus around 20 cm (8 in) long, which looks like a long, thick string with tufts of brownish-red wool attached to it (fig. 46c). At times, these tufts, which are masses of small intertwined branches, may completely obscure the heavily corticated main axis. This alga grows in the mid to high intertidal zone where it is commonly covered with diatoms (p. 2).

Distribution: *C. pikeanum*, Oregon border to Point Dume.

Ptilota (feathered)

Ptilota filicina is a bright red, finely branched plant with a distichously branched frond reminiscent of a fern (fig. 46a, 46b). It may be found on rocks from low tide levels into the shallow subtidal zone and is often abundant in drift weed in central and northern California. Plants are usually 15 to 30 cm (6−12 in) tall and are best recognized by using a hand lens to note that the ultimate branchlets are opposite, but that one member of each pair is quite large and the other very small. The larger ultimate branch has tiny marginal serrations. The related genus, *Neoptilota*, has a similar form but the small branch of each pair stops growing and never develops reproductive structures. A third feathery species, *Rhodoptilum plumosum*, has a dense fringe of short filaments along the margins of the main axes rather than small, serrated branches. All of these plants make beautiful pressings.

Distribution: *P. filicina*, *Neoptilota* spp., and *R. plumosum*, entire coast but rare south of Point Conception.

Phycodrys (oaklike seaweed)

The thick-veined blades of some species of *Phycodrys* resemble bright reddish-pink oak leaves (fig. 47a). The

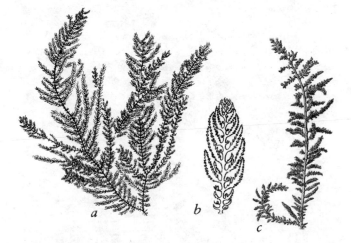

Fig. 46. *a*, *Ptilota filicina* (15 cm, 6 in); *b*, *P. filicina*, detail of branch (1 cm, 0.4 in); *c*, *Callithamnion pikeanum* (12 cm, 4.7 in)

most conspicuous species, *P. setchellii*, grows from the shaded low intertidal zone to depths of greater than 40 m (130 ft). Subtidal plants may reach lengths of 20 cm (8 in) and are particularly large and abundant in late summer. Veins of the smaller *Anisocladella pacifica* resemble those of *Phycodrys*, but the blades have toothed margins like *Nienburgia* (p. 143). *A pacifica* is commonly found around the rhizomes of surf grass (p. 161) in the sandy low intertidal zone.

Phycodrys setchellii was named for W. A. Setchell, professor of botany at the University of California at Berkeley from 1895 to 1934. Setchell, along with his associate, Nathanial Gardner (*Fryeella gardneri*, p. 132), made enormous contributions to algal taxonomy and to the marine flora of California.

Distribution: *P. setchellii* and *A. pacifica*, entire coast.

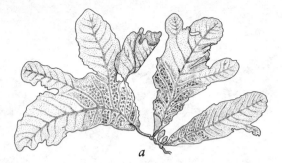

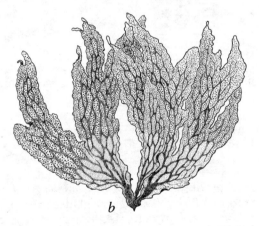

Fig. 47. *a, Phycodrys setchellii* (15 cm, 6 in); *b, Polyneura latissima* (12 cm, 4.7 in)

Polyneura (many veins)

This is a brightly colored red alga of lowermost intertidal rocks in somewhat sheltered situations. Extensive shallow subtidal populations contribute masses of drift material in some areas. The anastomosing veins of the crisp, membranous blades are unique and make it easy to identify. The only species is *Polyneura latissima* (fig.

47*b*), with delicate and often tattered and lacerated blades. Occasionally, perfect examples up to 30 cm (1 ft) tall may be collected and these make admirable mounted specimens. The reproductive bodies may be seen scattered as red spots over the surfaces of the blades.

Distribution: *P. latissima*, entire coast.

Nienburgia (after the German phycologist, Nienburg)

Nienburgia andersoniana is one of the interesting representatives of the algal family Delesseriaceae, which contains some of the most colorful and attractive of the membranous algae. This one consists of narrow, branched blades with a midrib in lower parts and with conspicuously toothed margins (fig. 48*b*). It varies considerably in width and stature, up to 30 cm (1 ft) tall in subtidal areas. We usually find smaller plants in shaded crannies and rock clefts, often growing with *Rhodymenia* (p. 126). It is sometimes protected by masses of sea grass leaves (p. 161). The plants resemble *Cryptopleura violacea* (p. 145) in color, size, and branching, but the marginal teeth, midrib, and symmetry of the blades are distinctive.

The majority of the Delesseriaceae are plants of deeper waters and some extend to great depths. Many are characterized by intricate vein patterns (e.g., *Polyneura*, p. 142) and some closely resemble the leaves of higher plants in this respect (e.g., *Phycodrys*, p. 140).

Distribution: *N. andersoniana*, entire coast.

Cryptopleura (hidden ribs)

Several common species of this genus occur in the low intertidal and subtidal zone in California. They are thin forms, some with long, narrow blades, others with ruffled, short blades, frequently bearing heavy midribs below and very fine, sometimes microscopic veins

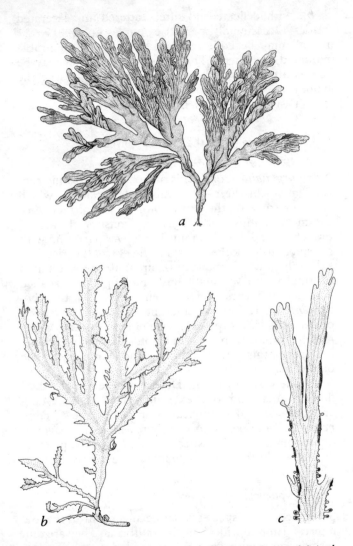

Fig. 48. *a*, *Hymenena flabelligera* (15 cm, 6 in); *b*, *Nienburgia andersoniana* (12 cm, 4.7 in); *c*, *Cryptopleura violacea*, upper portion of branch (7 cm, 2.8 in)

above. The margins of well-developed plants are often characterized by ruffles or small fertile bladelets. *Cryptopleura violacea* is a large species 10 to 20 cm (4–8 in) tall, deep rose to purplish-green and has few marginal proliferations but elongate tetrasporangial areas borne within the blade margins (fig. 48c). Free-living or epiphytic, *C. lobulifera* has very ruffled margins and is more purple than *C. violacea*. *C. corallinara* is a small, epiphytic species that is usually found attached to other algae, especially corallines and *Pterocladia* (p. 101).

Distribution: *C. violacea*, entire coast; *C. lobulifera*, entire coast but rare in southern California; *C. corallinara*, Monterey to Mexican border.

Hymenena (like a membrane)

These are thin, membranous plants up to 30 cm (1 ft) tall of the very low intertidal and subtidal zone in central and northern California. Their palmate or fan-shaped blades are divided into many segments and the microscopic or macroscopic veins usually diverge outwardly. Deep red *Hymenena flabelligera* is most common in the subtidal zone. Tetrasporic plants of this species are distinctive because the dark sporangia form slender patches in line with the upper veins (fig. 48a). Gametophytes and nonreproductive individuals resemble *Botryoglossum* (p. 116) but lack the small, dense ruffles and marginal proliferations common on the latter. *H. cuneifolia* has broad blades that may or may not have large marginal ruffles. *Hymenena* is generally smaller than *Botryoglossum* and lacks macroscopic veins in the upper portions. *H. multiloba* is a bushy, densely branched alga found on surf-exposed rocks. Tetrasporic plants are also distinctive in this species, the sporangia forming dark transverse bands across the narrow upper blades.

Distribution: *H. flabelligera* and *H. multiloba*, Oregon border to Pismo Beach; *H. cuneifolia*, Oregon border to Bodega Bay.

Myriogramme (countless lines)

Although related to *Hymenena* and other thin-bladed genera with veins, this genus lacks both a midrib and veins. *M. spectabilis* (fig. 49*a*) is the most common species in central California, but occurs only in subtidal habitats. The rose-red blades commonly occur as drift, however, and can usually be found after the first fall storms.

Distribution: *M. spectabilis*, Santa Cruz to Mexican border but rare south of Santa Barbara.

Acrosorium (summit sorus)

The single species is *Acrosorium uncinatum*, which is a warm-temperate plant that rarely occurs north of Santa Barbara. It is a curious alga, most often seen as an epiphyte and seemingly adapted to distribute and reproduce itself vegetatively by means of its small, hooked branches that cause it to become entangled and attached to various other algae (fig. 49*b*). The alga also grows on rocks and while floating free. Drifting masses over sand bottoms may reach abundances similar to algae attached to rocks. Although in Atlantic and Mediterranean regions it goes through a regular life history, in the Pacific it rarely shows anything but vegetative reproduction. The red color, narrow, branched, membranous blades with microscopic veins and frequent hooked branches distinguish it.

Distribution: *A. uncinatum*, Monterey to Mexican border.

Botryoglossum (grape tongue)

The large, deep red, slightly iridescent blades of *Botryoglossum farlowianum* are among the most beautiful of all large red algae (fig. 50). The plants are also among the most common seaweeds in northern and central California, growing in moderately exposed low intertidal habitats and into the subtidal zone. They are usually so abundant that plants can be found after only a brief

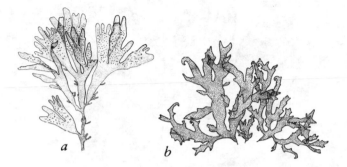

Fig. 49. *a, Myriogramme spectabilis* (12 cm, 4.7 in); *b, Acrosorium uncinatum* (6 cm, 2.4 in)

search of the drift in most locations. The alga may exceed 40 cm (16 in) in length and generally has densely ruffled margins with numerous proliferations. The less common *B. ruprechtianum* lacks ruffles, and the marginal proliferations on its flat, crisp blades are separated rather than densely overlapping.

A careful examination of the lower portions of *Botryoglossum* and *Hymenena* (p. 145) will occasionally reveal clumps of small pink to greenish blades growing on the surface. Except for their color, these blades resemble the marginal proliferations but are, in fact, another red alga, *Gonimophyllum skottsbergii.* This tiny, foliose plant is thought to be a parasite restricted to the above two host genera.

Distribution: *B. farlowianum, B. ruprechtianum,* and *G. skottsbergii,* entire coast, but rare south of Point Conception.

Tiffaniella (after the phycologist, L. Tiffany)

Beneath the protective cover of surf grass, sometimes on the sides and bottoms of shaded tide pools or in mixed mats of intertidal algae, one observes what appears like

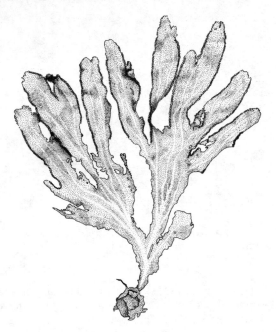

Fig. 50. *Botryoglossum farlowianum* (20 cm, 8 in)

fine, reddish-pink hair 2 to 5 cm (0.8–2 in) long. This may be dense enough to form red patches or tufts, or may be scattered among other small algae. These are the filaments of *Tiffaniella snyderiae*, which consists of a single branched row of elongate cells. It is distinctive in bearing asexual spores in polysporangia (fig. 51*c*), rather than tetrasporangia typical of most red algae.

   This is one of the plants cultivated by the remarkable nest-building ocean goldfish, the Garibaldi. The state marine fish, this territorial southern California species removes most other algae and sessile animals from 0.3- to 1 m- (1–3 ft) diameter areas on subtidal rocks, encouraging the growth of the red algae *Tiffaniella snyderiae*, *Pterosiphonia dendroidea* (p. 151), and

Murrayellopsis dawsonii. The latter species is rarely found outside areas under the care of the Garibaldi. Male fish do the gardening and the resulting dense red turf forms the nest upon which the female lays her eggs (pl. 7c). The eggs are apparently adapted for anchoring to these delicate algae, for they have extended filaments that wrap around the branches. These striking nests are best observed in the shallow subtidal zone during the July–August breeding season.

Distribution: *T. snyderiae*, entire coast; *M. dawsonii*, Monterey to Mexican border.

Polysiphonia (many tubes)

Polysiphonia is one of the familiar names among marine algae, for these plants have long been used in university biology classes as examples of the Rhodophyta. This has been so not only because they are common and widely available for instructive microscopic examination, but because *Polysiphonia* was the first red alga in which the life history was conclusively worked out (in 1906).

Its name refers to the multiple, often elongate and tubelike cells (pericentral cells) arranged in a cylinder around a central axial cell and, all being the same length, forming a series of "polysiphonous" segments.

There are several species of *Polysiphonia* of which a few, such as the tufted intertidal *P. paniculata* and the epiphytic *P. hendryi* are common. *P. pacifica* (fig. 51a) is also abundant and its many varieties are found throughout the intertidal and subtidal zones. The plants are usually small, mostly less than 10 cm (4 in) tall, very dark reddish to almost black in mass, but paler in subtidal habitats. The branches generally occur in whorls around the main axes and the tips are often provided with tufts of fine, colorless hairs. Although some of the larger species can be recognized by the use of a hand lens to detect the tiers of pericentral cells and terminal hair tufts, a microscope is needed for specific

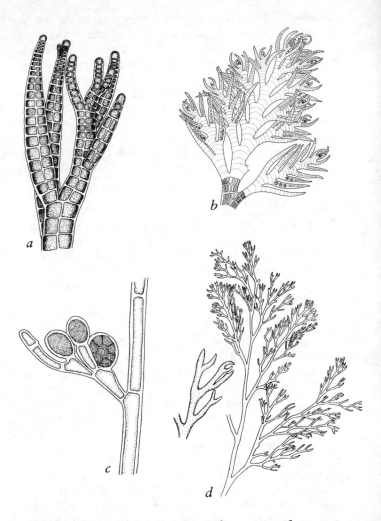

Fig. 51. *a, Polysiphonia pacifica* var. *pacifica*, upper branches (0.4 mm, 0.02 in); *b, Pterosiphonia dendroidea*, upper portion of plant (4 mm, 0.2 in); *c, Tiffaniella snyderiae*, vegetative filament and a branch bearing polysporangia (3 mm, 0.12 in); *d, Pterochondria woodii* (2 cm, 0.8 in)

identification. *Pterosiphonia* (p. 151) and *Pterochondria* (p. 151) are also polysiphonous but are distichously branched and lack hairs.

Distribution: *P. paniculata*, *P. hendryi*, and *P. pacifica*, entire coast.

Pterosiphonia (wing tube)

This is another delicate form closely related to *Polysiphonia* (p. 149) but more readily identified. The most common species, *Pterosiphonia dendroidea*, is a tufted, rock-inhabiting form whose branches often form low-lying mats. It is deep, dull red to almost blackish, 2.5 to 7.5 cm (1–3 in) long and consists of distichous, pinnate branchlets on a flattened axis. Individual main axes look like minute feathers (fig. 51b). The structure of the pericentral cells (pl. 10d) is comparable with *Polysiphonia*. *P. baileyi* is a larger, black, coarse form to 16 cm (6 in) tall with more widely spaced and less delicate branchlets.

Distribution: *P. dendroidea*, entire coast; *P. baileyi*, Crescent City to Mexican border.

Pterochondria (winged cartilage)

Pterochondria woodii (fig. 51d) is another common epiphyte, most abundant on *Cystoseira* (p. 90), but also found on *Macrocystis* (p. 73), *Nereocystis* (p. 70), and other large brown algae. It is a finely branched, distichous plant with regularly alternate branches but without percurrent axes. Except for the latter feature and its epiphytic habit, it somewhat suggests *Pterosiphonia* (p. 151). Male plants are distinctive in having small, terminal disks that bear the male gametangia.

Distribution: *P. woodii*, entire coast.

Chondria (cartilage)

Chondria californica is a distinctive member of this genus which may often be conspicuous in southern tide pools on sunny days because of a striking blue iridescence. It is a very slender, cylindrical plant that characteristically has tendrillike branch tips by which it entwines the branches of other algae and fastens itself. *Chondria* usually has a tuft of microscopic hairs at the tip of each branch. The dark red *C. nidifica* is a larger species that is usually found in sandy places on mid intertidal rocky shores where it is attached to stones slightly embedded in sand. It is erect, cylindrical, and fairly stiff (fig. 52b). Tetrasporangial plants are most distinctive, bearing dense tufts of very short lateral branchlets here and there on the axes.

   Chondria nidifica can resemble both *Neoagardhiella* (p. 120) and *Gracilaria* (p. 121), but the apical hairs and rather stiff branches aid in separating it from these latter two genera. In addition, only it and *Laurencia* (p. 154) host the parasite *Janczewskia* (p. 152).

   Distribution: *C. californica*, Santa Barbara to Mexican border; *C. nidifica*, Santa Cruz to Mexican border.

Janczewskia (after the French-Polish botanist, Janczewski)

The host of one of our most common parasitic red algae is *Chondria nidifica* (p. 152), which bears frequent little pink or whitish, burrlike growths on its axes and branches. These little burrs are fully developed plants of *Janczewskia lappacea* (fig. 52b), a species restricted to the one host. Other species grow on other species of *Chondria* and *Laurencia* (p. 154). These, like many parasitic algae, have reduced photosynthetic pigments and probably derive most of their sustenance from the host by haustorial cells that invade the host tissues. Although the vegetative plants are reduced, the full reproductive cycle is carried out through male, female, and tetrasporangial plants as in most other Rhodophyta.

   Distribution: *J. lappacea*, Morro Bay to Mexican border.

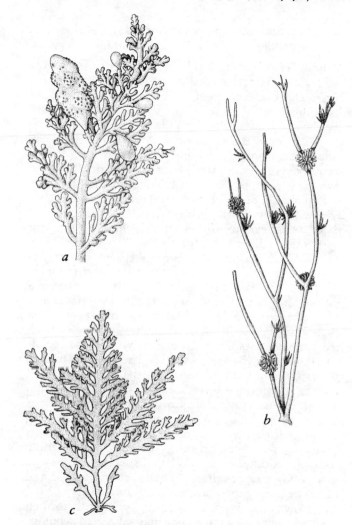

Fig. 52. *a*, *Laurencia pacifica* with epiphytic *Erythrocystis saccata* (6 cm, 2.4 in); *b*, *Chondria nidifica* with small, wartlike *Janczewskia lappacea* (15 cm, 6 in); *c*, *Laurencia spectabilis* (18 cm, 7 in)

Laurencia (after the French naturalist, de la Laurencie)

Most algae are recognized by their structural and morphological characters, and a few have distinctive textures. This one is frequently identifiable by smell, and its peculiar, acrid odor can be used to recognize it in the dark.

Laurencia is a genus of variable morphology. Examples exist both of cylindrical and flattened forms, frequently abundant throughout the intertidal zone. L. pacifica is perhaps the most common plant—a bushy, dark purplish, fleshy plant several centimeters tall, covered with very short, stubby branchlets that give it a papillate appearance (fig. 52a). It is the most frequent host of the bulbous red epiphyte, Erythrocystis (p. 155). L. spectabilis is a pinnate, flattened species of very different habit (fig. 52c) and with blades that resemble the leaves of sea rocket (p. 177). L. subopposita is a more openly branched form distinctive by the presence of hooked or entwining branches. This species is a common epiphyte in subtidal habitats, where its rose-red branches entangle with a variety of coarse red algae. All species, however, have a characteristic minute pit or slit at the tip of each branch, and this sunken growing point often bears a tuft of microscopic hairs.

It has recently been shown that the acrid odor of Laurencia mentioned above comes from halogenated compounds produced as secondary metabolites by this genus. These iodine-, bromine-, or chlorine-containing compounds differ from species to species and may be used as taxonomic criteria. The function of these and other secondary metabolites found in a variety of red and brown algae is presently unknown, but their often toxic nature suggests that they may serve to discourage epiphytes and grazers. If this is the case, then the California sea hare, a marine molluscan grazer resembling a large slug, has adapted to this chemical system, for its larvae will settle and develop on Laurencia pacifica. Moreover, a variety of sea hares feed on algae containing various toxic substances and concentrate the chemicals in their digestive glands. The general lack of predation on these animals indicates they may be using the algae as sources of both energy and defensive compounds. The California sea hare does use the phycobilin pigments in its red algal food to derive a defensive "ink" that it releases when disturbed. If fed on algae lacking these pigments, ink production stops. In addition to aiding sea hares, secondary metabolites of algae may eventually aid man, as some are showing promise as potential antibiotics and insecticides.

Distribution: *L. pacifica* and *L. subopposita*, Monterey to Mexican border; *L. spectabilis*, entire coast.

Erythrocystis (red bladder)

One of the best clues to the identification of *Laurencia* (p. 154), especially *L. pacifica*, is the presence of a small, reddish-purple, balloonlike, epiphytic plant which grows from the apical pit of some of the branches (fig. 52*a*). This epiphytic red alga, *Erythrocystis saccata*, is usually about 0.5 to 3 cm (0.2–1.2 in) high of simple, saccate shape. Several plants may arise from one pit. This lovely little alga is best developed in late summer and early fall.

Distribution: *E. saccata*, Monterey to Mexican border.

Rhodomela (red to black)

This is another red alga that ranges from central California north to the Bering Sea. *Rhodomela larix* (fig. 53*b*) consists of several wiry axes to about 20 cm (8 in) long, closely beset with spirally arranged clusters of cylindrical branchlets about 0.5 cm (0.2 in) long. These brownish-black, ropelike plants are characteristic inhabitants of mid intertidal rocks and sometimes form a dominant part of the vegetation. They also harbor the warty brown algal epiphyte, *Soranthera* (p. 62).

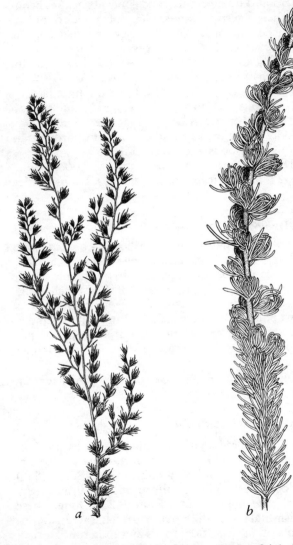

Fig. 53. *a, Odonthalia floccosa* (20 cm, 8 in); *b, Rho-domela larix* (10 cm, 4 in)

Distribution: *R. larix*, Oregon border to Point Conception.

Odonthalia (tooth branch)

This is a genus characteristic of Canadian and Alaskan shores. *Odonthalia floccosa* extends southward and is well represented in northern California. It reaches 45 cm (1.5 ft) in height and is blackish-brown. Major branches are slender and slightly flattened. They bear alternate, distichous short branchlets that, in turn, bear short, flat, and pointed ultimate branchlets. The latter are usually clustered in groups (fig. 53*a*). Tattered *O. floccosa* must be carefully examined to distinguish them from *Rhodomela* (p. 155).

Distribution: *O. floccosa*, Oregon border to Point Conception.

6

Sea Grasses

Although the vast majority of seashore plants are algae, there are many coastal localities at which at least one kind of flowering plant can be found growing under strictly marine conditions in intertidal habitats or below. Sometimes these sea grasses may be so abundant as to form extensive beds on sand, mud, or rocks to the virtual exclusion of other kinds of plants. Only two genera of sea grasses occur in California, and they may be identified by the illustrations and notes that follow.

Zostera (girdle or band) Eel Grass

The true eel grass, *Zostera marina*, is a plant of quiet waters. It commonly lives on tidal mud flats and in bays and estuaries from low tide level down to 6 m (20 ft) or more. In these habitats, its dull, light-green blades often form extensive beds that provide shelter for numerous marine animals and its roots stabilize the muddy substratum. The illustrations in figure 54 show three variants of eel grass; those on the left represent narrow-leaved forms characteristic of sheltered bays and salt marshes (fig. 56). The large, broadleaf form shown on the right is known as var. *latifolia* and occurs on sheltered sandy bottoms along the open coast such as at La Jolla Bay. These open-coast populations have been observed at depths of 30 m (100 ft) and after storms, frag-

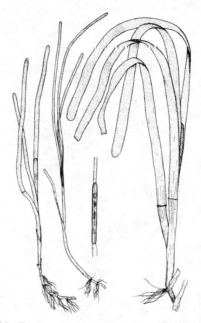

Fig. 54. *Zostera marina*, showing three leaf forms (30 cm, 12 in) and part of a fertile stem (5 cm, 2 in)

ments are commonly cast up on sandy beaches. Observations of *Zostera* growing in relatively clear water and experiments with artificial shades indicate that light is the major factor limiting subtidal growth in the generally turbid water of bays and estuaries.

The flowers and fruits of eel grass are rather obscure. Part of a fertile stem is shown in figure 54, and two developing seeds are visible where the enveloping spathe does not completely cover them. *Zostera* seeds are important food for many waterfowl, particularly black brant. In addition, they are one of the few marine seeds used by man, being harvested by the Seri Indians in the Gulf of California and used like wheat. Live vegetative portions of the plant are eaten by sea turtles and

Fig. 55. *Phyllospadix torreyi* with flowering stems (40 cm, 16 in)

some marine invertebrates, and dead material lost through senescence can provide vast amounts of detritus that is utilized by a variety of animals. Because of the great importance of eel grass to coastal environments, it received much attention following its almost complete disappearance from both sides of the Atlantic in the early 1930s. Although the loss of productivity was undoubtedly important, the most immediate effect was erosion of the substratum. This destructive "wasting disease" of eel grass was probably caused by unusually high water temperatures. It took fifteen years for the plants to return to normal growth, and there has not been another serious decline since.

Distribution: *Z. marina*, Oregon border to San Diego.

Phyllospadix (leaf inflorescence) Surf Grass

Two species of surf grass occur widely along the Pacific coast. Unlike *Zostera* (above), these plants normally grow along the open coast from the low intertidal zone down as deep as 6 m (20 ft). *P. torreyi* or Torrey's surf grass (fig. 55) has narrow (1–2 mm [0.04–0.08 in]) somewhat wiry leaves that are elliptical in cross section and long, flowering stems bearing several spadices as shown in the figure. It is most common in protected sandy areas. *P. scouleri* (Scouler's surf grass) has broad (2–4 mm), flat leaves and short, basal flowering stems bearing only one or two spadices (pl. 11*a*). It is most abundant attached to rocks in areas of heavy surf. Both species are abundant intertidally in California, forming extensive emerald green masses near mean low tide (pls. 3*a*, 3*b*), where they are commonly mistaken by the layman for eel grass.

The seaweed observer should not overlook the importance of surf grass as a habitat for various kinds of algae. Not only do epiphytic species occur regularly on it (*Melobesia*, p. 108; *Smithora*, p. 98), but others are hidden beneath the protecting layer of leaves. By simply spreading and opening up the mantle of leaves to reveal the inhabitants under them, you will find many species otherwise passed over unseen. The plant also provides habitat and food for many marine animals. The protected bases are a favored environment for juvenile spiny lobsters in southern California; and the small, laterally compressed limpet, *Notoacmea paleacea*, is found only on the blades of *Phyllospadix*. Unfortunately, the blades are particularly susceptible to damage by oil pollution. As a result of the Santa Barbara oil spill, tons of blades were lost. This loss of habitat probably also affected the other members of the surf grass association mentioned above. Fortunately, the rootstocks and basal meristems survived the pollution to produce new blades.

Distribution: *P. torreyi* and *P. scouleri*, entire coast.

7

Coastal Salt
Marsh Vegetation

In addition to the sea grasses that are regularly and completely submerged by sea water, there are seashore plants referred to as halophytes, or salt plants, which live in marine marshes or along the coastal strand under the influence of high soil salinities, salt water inundation, or a combination of these factors. In marine marshes, these plants are only partially or occasionally inundated by the sea. Some of them are algae discussed in previous sections, but the most conspicuous are flowering plants adapted to growth in salty soil. Several interesting forms grow along the California coast in marshes adjoining San Diego Bay, Morro Bay, Moss Landing, San Francisco Bay, Tomales Bay, Humboldt Bay, and other coastal areas.

Depending on their tolerance for inundation, amount of salt in the soil, and other physical and chemical factors associated with a saltwater marsh, and their ability to compete with associated plants for space and light, these halophytes often are found in particular areas of the marsh just as algae are found in particular areas or zones in the open coast rocky intertidal. A typical zonation pattern of the most common marsh plants is shown in figure 56. It is curious that most if not all of the flowering plants in salt marshes are not obligate halophytes but, in greenhouse culture, will grow in fresh water and often even grow best under these conditions. Apparently their adaptations to deal with the physical and chemical environment discussed above allow them to tolerate and grow in salt marshes, but their poor competitive abilities against strictly freshwater or terrestrial plants prevent them from growing in other areas. A few of the most common species are discussed below, beginning with plants of the lowest tidal levels and proceeding to those found most often in the high marsh.

Spartina (cord) Cord Grass

The stout, bushy, coarse grass of the low marsh (below the +1.5 m [5 ft] tidal level; fig. 56), with leaves up to 1 cm (0.4 in) broad at the base, is the California cord grass, *Spartina foliosa* (pls. 11*b*, 11*d*). This is a plant up to 1 m (3.3 ft) tall, of marsh waterways in which the tide often covers half or more of its length. Thus, at high water, *Spartina* stands partially submerged like rice in a paddy. The leafy stems arise from extensive creeping rhizomes buried in the salty mud. The inflorescences are dense, spikelike structures 15 to 20 cm (6–8 in) long.

The mud beneath and between California cord grass stems is commonly covered with various algae, including films of golden-colored, unicellular diatoms (p. 2), the brownish-green *Enteromorpha* (p. 51), red *Polysiphonia* (p. 149), the brownish-green, siphonous *Vaucheria* (p. 51), and blue-green algae (p. 3; pl. 11*d*). All of these kinds of algae can be important contributors to marsh productivity and, in addition, some of the blue-greens can convert nitrogen gas into other nitrogen-containing compounds that are important nutrients for the algae as well as the flowering plants.

Although locally common (e.g., south San Francisco Bay, Humboldt Bay, Tijuana River), California cord grass is not abundant in most Pacific coast salt marshes and is absent in many areas such as Bodega Harbor, Tomales Bay, Elkhorn Slough, Morro Bay, and Goleta Slough. This situation is quite different from the Atlantic coast of the United States, where another spe-

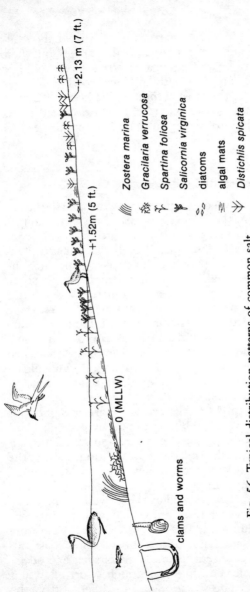

+2.13 m (7 ft.)

+1.52 m (5 ft.)

0 (MLLW)

clams and worms

Zostera marina	
Gracilaria verrucosa	
Spartina foliosa	
Salicornia virginica	
diatoms	
algal mats	
Distichlis spicata	
Frankenia grandifolia	

Fig. 56. Typical distribution patterns of common salt marsh plants

cies, *Spartina alterniflora*, is usually the most common marsh plant, producing vast amounts of food used by animals in the marsh and adjacent environments. These extensive stands of cord grass are sometimes cut for fodder, and in some areas livestock use the marsh directly as grazing land.

All marsh plants face the problem of high levels of salt, both in the soil and in the water. Among other things, this makes it difficult for them to extract and hold the fresh water necessary for maintenance and growth. A variety of mechanisms are found in marsh plants to overcome this problem. *Spartina* and a few other genera have epidermal salt glands that excrete excess salt. After evaporation, these excretions are evident as small white patches of salt crystals on dry leaves. Aerenchyma tissue is also present in the stems, allowing oxygen transport to the roots that generally grow in waterlogged anaerobic mud.

Distribution: *S. foliosa*, Humboldt Bay to Mexican border.

Salicornia (salt horn) Pickle Weed, Glasswort

The most distinctive of our common halophytes is *Salicornia*, which has peculiar, succulent, jointed stems. Annual and perennial species occur, but they are similar in appearance and rather difficult to identify specifically because of their exceedingly obscure flowers that are sunken in the fleshy axes. *S. virginica* (*pacifica*), Common pickle weed, is the most abundant perennial species (pls. 11*b*, 11*c*) and Bigelow's glasswort (*S. bigelovii*) is the most common annual. Both species occur from mean high water to the highest reaches of the marsh (fig. 56).

Pickle weed has had a long tradition of utilization in Europe as a fresh or a pickled vegetable. So palatable are some of the species that in certain of the older horticultural works it was recommended that they be cultivated as a vegetable by imitating a portion of the salt marsh. Owing to their high yield of soda, several species were formerly used in making glass and soap, the ashes of the glasswort being known in the trade as barilla.

Salicornia solves the salt problem by storing large amounts of water in its succulent stems and excess salt in its tissue. At the end of the midsummer growing season in central and northern California, these stems turn red and eventually dry up and fall off the plant, carrying away the accumulated salt and contributing to the detritus in the marsh. Stands of *Salicornia* also often appear orange from a distance, the result of being covered with the small orange stems of salt marsh or alkali dodder, *Cuscuta salina*. This genus, which is parasitic on a variety of terrestrial plants, penetrates and extracts material from its host by minute haustoria. Although a true plant with small white flowers, dodder lacks chlorophyll and has lost the ability to produce its own food. As the stems spread through the *Salicornia*, they give the marsh the appearance of being covered with masses of orange string (pl. 11c).

Distribution: *S. virginica*, entire coast; *S. bigelovii*, Santa Barbara to Mexican border; *C. salina*, entire coast.

Jaumea (after the French botanist, I. H. Jaume)

Frequently growing in mid-marsh with Common pickle weed (above) is the somewhat succulent, perennial herb, *Jaumea carnosa* (fig. 57b). Like many marsh species, the green, upright portions of fleshy *Jaumea* arise from a system of prostrate stems left from the previous year's growth. Most marsh plants lack large, colored flowers, but *Jaumea* is the exception. Its yellow disk flowers with distinctive fleshy involucres are most common from May through October.

Distribution: *J. carnosa*, entire coast.

Batis (seashore plant) Saltwort

Batis maritima, the only species in this genus, can be found throughout southern California marshes as well

Fig. 57. *a, Triglochin maritima*, entire plant, leaves, terminal spike, and flowers (entire plant 50 cm or 20 in tall); *b, Jaumea carnosa* (20 cm, 8 in); *c, Batis maritima*, portion of plant (10 cm, 4 in), flower, and fruit

as along the coastal strand. The generally prostrate or ascending stems may become woody at the base. They are up to 1 m (3.3 ft) long and bear opposite, fleshy leaves. The separate male and female flowers occur in distinctive catkins (fig. 57c).

Distribution: *B. maritima*, Morro Bay to Mexican border.

Triglochin (three point) Arrow Grass

When in flower, seaside arrow grass *Triglochin maritimum* can be mistaken for no other plant in the marsh, for its flower stalks reach lengths of more than 70 cm (2.3 ft; fig. 57a). These stalks rise from a base of densely tufted, narrow leaves above a thick rootstock. Two other species of *Triglochin* also occur in coastal marshes, but are relatively rare. Although commonly found in the high marsh (above the +1.5 m [5 ft] tidal level), arrow grass, like many other marsh plants, is also found inland on alkaline flats or in other saline habitats.

Distribution: *T. maritimum*, entire coast.

Suaeda (arabic name) Sea Blite, Seep Weed

The common marsh *Suaeda* in central and southern California is *S. californica* (fig. 58a), or California sea blite. It is a fleshy, dull grey, decumbent to semierect bush 1 m (3.3 ft) or more wide and up to 1 m tall. The leaves are simple and linear, about 1.3 cm (0.5 in) long and usually densely clothed with minute downy hairs. Its sometimes reddish stems are usually woody near the base. The greenish flowers are small, clustered, and obscure without any colored petals. Several other species of sea blite occur in California marshes, but most inhabit inland alkaline flats.

Distribution: *S. californica*, San Francisco to Mexican border.

Limonium (marsh) Sea Lavender, Marsh Rosemary

This is a large genus of perennial herbs with a rosette of basal leaves and tall, branched flower stalks. Three spe-

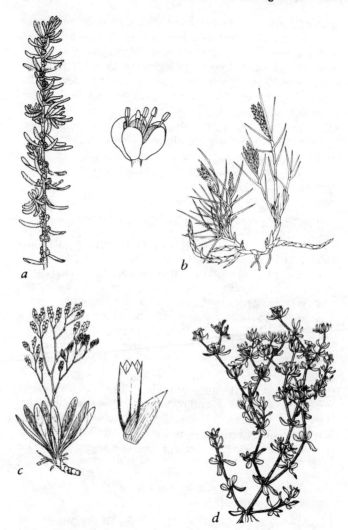

Fig. 58. *a*, *Suaeda californica*, plant (15 cm, 6 in) and flower; *b*, *Distichlis spicata* (15 cm, 6 in); *c*, *Limonium californicum*, plant (80 cm, 31 in) and flower; *d*, *Frankenia grandifolia* (30 cm, 12 in)

cies occur in California, two of which are relatively rare introductions that escaped from cultivation. Our native and common salt marsh species, *Limonium californicum*, produces flower stalks to 50 cm (20 in) tall which bear flowers with white sepals and violet petals (fig. 58c). The sepals are very membranous and some European species are used in dry flower arrangements. Like cord grass (p. 163), California marsh rosemary has salt glands as well as specialized air-conducting tissue that aids in supplying oxygen to the roots.

Distribution: *L. californicum*, Eureka to Mexican border.

Distichlis (two ranked) Salt Grass

The most common halophytic grass along salt marshes and sandy flats of California is *Distichlis spicata* (fig. 58b). Salt grass is a perennial plant consisting of extensive, creeping, and sometimes underground, scaly rhizomes from which stiff, harsh, and somewhat spiny leaves arise to form dense stands. Barefoot beach goers remember the plant well after having the painful experience of walking on it. It tends to grow at the upper margins of marshes or above the normal high tide line on the shore, where only unusually high water may cover it. Like cord grass (p. 163), *Distichlis* excretes excess salt that may be seen (and tasted) as white specks on the leaves. The plants are ordinarily low and spreading, but may sometimes bunch up to 30 cm (1 ft) high.

Distribution: *D. spicata*, entire coast.

Frankenia (after the Swedish botanist, J. Franke) Alkali Heath

Frankenia grandifolia (fig. 58d) is one of the few abundant plants of our salt marshes that bear noticeable flowers, but even these are only 1.5 cm (0.6 in) across. When in full bloom, however, they are an attractive pink and lend a modicum of bright color to the varied shades of green on the marsh. *F. grandifolia* is a low bush 10 to 40 cm (4−16 in) tall, bearing short, narrow leaves in groups. *F. palmeri*, sometimes called Palmer's Frankenia, is a smaller southern California species with even smaller leaves (less than 4 mm or 0.15 in long) and white flowers. This genus, like salt grass (p. 170), also inhabits the upper marsh margins where the soil is infrequently covered with salt water.

Distribution: *F. grandifolia*, Bodega Bay to Mexican border; *F. palmeri*, San Diego to Mexican border.

Monanthochloe (one flower) Salt Cedargrass, Salt Flat Grass

Monanthochloe littoralis is another marsh grass commonly associated with salt grass (p. 170). This perennial species has creeping, very wiry stems bearing clusters of short, curved leaves (fig. 59*a*). It is common in salt marshes throughout the southern United States and extends into Mexico and Cuba.

Distribution: *M. littoralis*, Point Conception to Mexican border.

Cressa (Greek, Cretan woman) Alkali Weed

The grayish, woolly haired *Cressa truxillensis* is found throughout California in alkaline areas but also occurs in southern California coastal salt marshes. Growing in the high marsh, these 10 to 20 cm (4−8 in) tall, tufted plants (fig. 59*c*) bear small white flowers in the leaf axils from May to October.

Distribution: *C. truxillensis*, Santa Barbara to Mexican border.

Scirpus (rush) Bulrush, Tule

Scirpus is a large sedge genus of primarily fresh water marsh plants characterized by generally triangular or oval upright stems that rise from thick rootstocks and

Fig. 59. *a, Monanthochloe littoralis* (20 cm, 8 in); *b, Scirpus californicus* (plants to 4 m or 13 ft tall); *c, Cressa truxillensis* (20 cm, 8 in)

rhizomes. In the larger species, these dark green stems may reach lengths of more than 4 m (13 ft) and form dense tule thickets. They invade salt water marshes from the fresh water end and cannot tolerate high soil salinities. *S. californicus* (fig. 59*b*) or California Bulrush is common in north San Francisco Bay, where it occurs seaward as far as Martinez and along fresh water streams running into San Pablo Bay. Indians used its buoyant stems to construct rafts. California bulrush has triangular upper stems, while those of the very similar *S. acutus* are oval or cylindrical throughout. *Juncus acutus* or spiny rush, a true rush with worldwide distribution, is also found in coastal marshes. Much smaller than the tules above, it forms large perennial tussocks generally less than 1 m (3.3 ft) tall. In other parts of the world, *Juncus* species are cut for cattle and sheep fodder but the tough, prickly leaves of *J. acutus* are eaten only by camels.

Distribution: *S. californicus*, San Francisco to Mexican border; *S. acutus*, Oregon border to San Diego; *J. acutus*, Morro Bay to Mexican border.

8

Coastal Dune
Vegetation

While visiting the shore to observe seaweeds, it is difficult not to notice the various flowering plants characteristically found on the slightly elevated terrestrial margins of the sea. This region of the coast, known as the coastal strand, includes environments not generally subject to inundation by sea water as are the marine areas proper and marshes; but these areas are nonetheless influenced by the adjoining sea via storm waves, wind, salt spray, and blowing sand. The coastal strand includes a variety of habitat types, but we will describe here only the most common species of plants found from the beach to the top of the foredune, that part of strand extending from the highest tide level on the shore to the highest point on the sea side of the dunes and thus under the greatest influence of the sea. Many of the plants found here also extend into the backdune, mixed with others that are found in other completely terrestrial habitats. Some of these foredune plants are also found in other coastal strand habitats; very few beach plants are restricted to the beach alone.

The general distribution of some of the most common beach and foredune plants in central California is shown in figure 60. Although the figure suggests a zonation of species in this area, this feature is not nearly as evident as it is for plants in the rocky intertidal zone or salt marshes. Tidal inundation, the major climatic factor in the latter two environments, varies dramatically over very short vertical distances. Although salt water occasionally covers the beach, tidal inundation does not directly affect the foredune, and other environmental factors have relatively broad vertical gradients and also vary horizontally with local differences in topography. Thus, although there are general distributional trends, one can find almost any plant described below at any place on the foredune.

In addition to wind, salt spray, and a rather mobile substratum, dune plants also have to contend with very low soil nitrogen and a general lack of surface soil water. Except for the high tide drift line where nutrients are released from decomposing organic matter, litter is scarce and nitrates are easily leached out of the porous sand. In compensation for this, many of the plants have developed symbiotic relationships, both with nitrogen-fixing bacteria and with fungi. The fungi do not fix nitrogen, but form mycorrhizal relationships with the plants whereby the plant receives nutrients and water from the extensive underground hyphae of the fungus as it decomposes organic matter, and the fungus gets food from the plant. A sand dune is probably the last place most of us would look for mushrooms, but they are commonly present on the backdune after heavy rains. Plant-fungal associations are also found on the foredune, but the fungi are types that usually do not produce mushrooms.

Water drains rapidly through the porous sand and, except during the rainy season, the plants must rely on fog, dew, fungal transfer, storage in succulent tissue, and/or tapping the subsurface fresh water table for fresh water. As a result of the excellent soil drainage, soil salinity is surprisingly low except on the beach after storm waves and does not appear to be a major factor affecting the vegetation. Recent research on the growth of crop plants on sand watered with salt water utilizes the drainage characteristics of dunes. Through extensive breeding, barley and tomato varieties have been selected which can yield reasonable harvests when grown

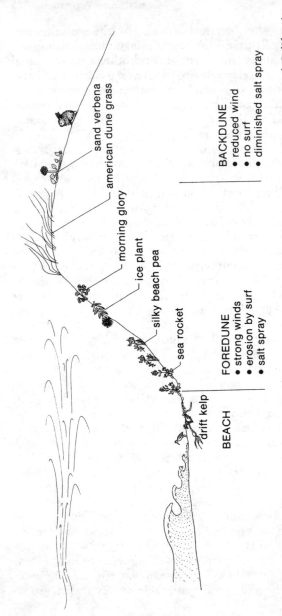

Fig. 60. Profile of beach and foredune showing the distribution of the most common plants in central California

BACKDUNE
- reduced wind
- no surf
- diminished salt spray

FOREDUNE
- strong winds
- erosion by surf
- salt spray

BEACH

sand verbena
american dune grass
morning glory
ice plant
silky beach pea
sea rocket
drift kelp

under such conditions on dunes. The plants cannot survive high soil salinities, and the well-drained sand prevents such a salt build-up. This method of growing crop plants has the potential of revolutionizing agriculture in arid countries with extensive sandy coasts.

Cakile (Arabic name) Sea Rocket

Sea rocket gets its name from the distinctive seed capsules that resemble a two-stage rocket (fig. 61a). When mature, the upper "stage" falls from the plant and, because the fruit is corky, it may float for many days in salt water. If during this ocean voyage it happens to land on the upper beach and is washed free of salt, germination readily occurs. The lower "stage" is buried with the dead or dying annual vegetative portions of the plant and can germinate in place. Thus, sea rocket has the ability to disperse as well as the ability to continue occupation of suitable habitats.

The most common species in California is *Cakile maritima* (fig. 61a; pls. 12a, 12e), introduced to the state in the early 1930s. It often grows closer to the sea than any other beach plant and is occasionally covered with salt water when high tides and strong onshore winds combine to push the sea high on the shore. The succulent stems and leaves of this annual aid in storing water and its generally low-growing habit probably helps resist sand abrasion. The purple to pink flowers resemble those of the related wild radish found in disturbed terrestrial habitats, and its pinnately cleft leaves are shaped like the thallus of the red alga, *Laurencia spectabilis* (p. 154). The other introduced but less common sea rocket, *C. edentula*, has dentate leaves and smaller petals (less than 6 mm or 0.2 in long), pinkish-white flowers and a different-shaped fruit.

Although naturally most common on the foredune, *Cakile maritima* will grow well even in grasslands if the surrounding vegetation is removed, and labora-

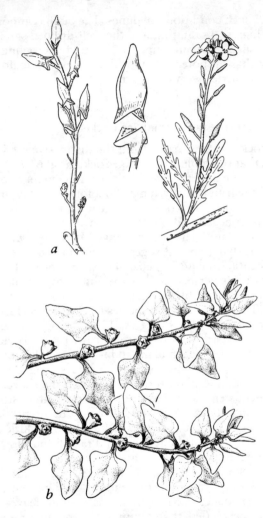

Fig. 61. *a, Cakile maritima,* fruiting stem, fruit, and flowering stem (10 cm, 4 in); *b, Tetragonia tetragonioides,* portion of stems (20 cm, 8 in)

tory and field research suggest that grassland species shade out the relatively slow-growing sea rocket that has a very high light requirement. So, like most beach and salt marsh plants, *C. maritima* is not an obligate halophyte, but lives on the foredune because few other plants can compete with it there.

Distribution: *C. maritima*, entire coast; *C. edentula*, Oregon border to Fort Bragg.

Tetragonia (four angles) Sea Spinach

New Zealand spinach (*Tetragonia tetragonioides*) is a spreading annual introduced from the southern hemisphere where it is commonly eaten as a vegetable. Like sea rocket above, it occurs from the upper beach (pl. 12*a*) to the top of the foredune and is extremely resistant to salt spray. The fleshy, dark green, triangular leaves and thick stems (fig. 61*b*) are covered with small, crystalline swellings that give the plants a crisp appearance. The yellow-green flowers, located in the leaf axils and generally covered by the leaves, lack petals and are rather obscure.

Distribution: *T. tetragonioides*, Oregon border to Los Angeles.

Lathyrus (pea) Pea

The delicate purple and white flowers of the silky beach pea, *Lathyrus littoralis*, are often mixed with those of sea rocket (p. 177), for the two commonly grow together on the low foredune. Befitting its common name, this generally low-growing, perennial legume is covered with fine silky hairs. The gray-green, compound leaves have four to eight closely spaced leaflets and the "pea pods" formed during the April to July flowering period contain one to five seeds (fig. 62*a*; pl. 12*c*).

Distribution: *L. littoralis*, Oregon border to Point Conception.

Fig. 62. *a, Lathyrus littoralis*, flower, branch (15 cm, 6 in), and fruit; *b, Mesembryanthemum crystallinum* (15 cm, 6 in); *c, M. edule* (12 cm, 4.7 in)

Abronia (graceful) Sand Verbena

Perhaps the most colorful of the dune plants are the sand verbenas, of which there are three common kinds: the crimson-flowered perennial red sand verbena (*Abronia maritima*; fig. 63*a*), which grows on the lower foredune, the rose-flowered annual beach or common sand verbena (*A. umbellata*; fig. 63*b*), most abundant on the high foredune and backdune, and the yellow-flowered perennial yellow sand verbena (*A. latifolia*), found on the beach and throughout the dunes. These are all prostrate herbs with opposite entire leaves and rather thick, succulent stems. The roots are usually stout and fleshy, deeply embedded in the sand. The most characteristic feature of the herbage is its covering of minute glandular hairs that exude a sticky material to which grains of sand adhere. Thus, the plants appear as if sprinkled with sand, but one finds that the sand does not shake off. The flowers are fragrant and are borne in dense clusters. Like many beach species, the flowering season is long, extending from February to October or November.

Distribution: *A. maritima*, near Big Sur to Mexican border; *A. umbellata*, Bodega Bay to Los Angeles; *A. latifolia*, Oregon border to Santa Barbara.

Mesembryanthemum (midday flower) Ice Plant, Sea Fig, Hottentot Fig

This is a very large genus of succulent plants native to the southern hemisphere, particularly South Africa, where hundreds of species provide some of the most colorful natural ground covers in the world. Because of their abundant flowers, often of striking brilliance, the ice plants have long been popular garden plants in the warm coastal regions of California. They have also been used widely as sand-stabilizing and erosion-control plants wherever a drought-resistant succulent can be used. From these plantings, the ice plants have escaped and become naturalized along the California shore. Unfortunately, some of them are especially well adapted to compete with the native vegetation of the coastal strand, so they are frequently seen spreading over upper beach sands and over dunes.

The true ice plant, called common ice plant, is *Mesembryanthemum crystallinum* (fig. 62*b*), named for the very large vesicular cells covering the surface of stems and leaves, which give it the sparkling appearance of being adorned with globules of ice. The plant is an annual with broad, oval leaves that are green and very succulent in spring. In summer, the older plants become reddish and the leaves reduced. The flowers are white or pinkish and not especially striking.

The more conspicuous succulents that are generally called ice plants, but more correctly sea fig or Hottentot fig, are two species with elongate, succulent leaves of triangular shape in cross section. These are the creeping, deep green, fleshy plants so widely used as roadside binders and in erosion control on dry slopes. The plants become so heavy with stored water that they have been known to slide off steep, wet slopes during heavy rains. The sea fig (*Mesembryanthemum chilense*) has small (less than 5 cm [2 in] in diameter) magenta-colored flowers and the Hottentot fig (*M. edule*) has larger yellow flowers (fig. 62*c*). The former is from South America and the latter from South Africa. *M. edule* was a traditional food plant of the Hottentots, since the fruits are edible. Along the California coast, the fruits, along with the leavings of shore visitors, support large populations of ground squirrels. The crushed leaves are said to act as a deodorant when rubbed on the body.

Several other species have become naturalized on sea bluffs and low ground along the shore. Although all have succulent leaves, the leaves of different species are unlike in form and appearance. The flowers, however, whether white, rose, purple, or yellow, are all similar in

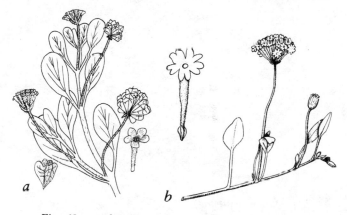

Fig. 63. *a, Abronia maritima,* fruit, portion of flowering stem (15 cm, 6 in), and flower; *b, A. umbellata,* flower and portion of stem (15 cm, 6 in)

their wheellike shape, numerous slender petals, and abundant stamens.

Distribution: *M. crystallinum,* Monterey to Mexican border; *M. chilense* and *M. edule,* entire coast.

Calystegia (cup covering; formerly *Convolvulus*) Morning Glory

The beach morning glory (*Calystegia soldanella*) has large, distinctive pinkish or purplish flowers (pl. 12c). It blooms from April to August along the coast, growing perennially from deep-seated rootstocks and producing fleshy, prostrate stems up to 60 cm (2 ft) long from the root crown. The leaves are kidney shaped, thick, and somewhat fleshy. Not only is it found along the California coast, but beach morning glory extends north to British Columbia and occurs in South America and on many other Pacific sea shores.

Distribution: *C. soldanella,* Oregon border to San Diego.

Fig. 64. *Atriplex leucophylla* (15 cm, 6 in)

Atriplex (salt bush) Salt Bush

The salt bushes are characteristic plants of alkaline sinks, dry lakes, and saline areas throughout the west. Several species occur in our coastal salt marshes and others grow along beaches and on the coastal strand. A common one is *Atriplex leucophylla* or beach salt bush (fig. 64). Some of the inland salt bushes and weedy forms that appear in waste places are naturalized from Eurasia and from Australia.

These are mostly dull-looking shrubs of grayish or whitish color, covered with small bladder cells used for salt removal. They bear separate male and female flowers that are obscure, small structures without colored parts. There is nothing glamorous or striking about the salt bushes, but they are often a prevalent component of the dune vegetation.

Distribution: *A. leucophylla*, Fort Bragg to Mexican border.

Ambrosia (food of the gods) Ragweed

Ambrosia chamissonis is a common coastal species in the ragweed genus, but is called beach burr or silver

Fig. 65. *Ambrosia chamissonis* (15 cm, 6 in)

beachweed. Its pollen is not noted for causing an allergic reaction, but the spine-covered fruits can be the bane of barefoot sunbathers and beach volleyball players. Silver beachweed is a low-growing, perennial herb whose silvery, silky-haired leaves and silver to reddish stems form extensive, loose mats over the sand. The leaves are of highly variable form and range from simple ovals to pinnately cleft shapes resembling sea rocket (p. 177). The male flowers occur in terminal spikes with the spiny females below (fig. 65). The latter develop into brown burrs as the fruit matures.

Distribution: *A. chamissonis*, entire coast.

Camissonia (nightgown; formerly *Oenothera*) Evening Primrose

The evening primrose of California beaches and dunes is the yellow-flowered *Camissonia cheiranthifolia*, commonly called beach primrose. This common name is more appropriate than evening primrose both be-

Fig. 66. *Camissonia cheiranthifolia* (20 cm, 8 in)

cause of the plant's habitat and because its flowers open in the morning rather than evening. Beach primrose has silvery, downy foliage that consists of a basal rosette of elongate leaves from which several prostrate stems radiate for up to 0.5 m (20 in). The flowers have bright yellow petals about 1 cm (0.4 in) long in central and northern California and up to 2.5 cm (1 in) in southern California. They sometimes have reddish spots. Flowering begins in the spring and continues throughout the summer. The fruit capsules become peculiarly coiled at maturity (fig. 66).

Distribution: *C. cheiranthifolia*, entire coast.

Ammophila (sand loving) Marram Grass, Beach Grass

Ammophila arenaria (pl. 12*b*), commonly called European beach grass, Holland dune grass or marram grass, is native to Europe, but has been introduced into North America, South Africa, and New Zealand in efforts to stabilize sand dunes. Its pale-green leaves may reach heights of more than 1 m (3.3 ft) and form dense stands or tussocks on central and northern California coastal dunes.

Marram grass is perfectly adapted to life on the shifting sands of the foredune, where it reaches its greatest development. The plant spreads rapidly by underground rhizomes that send vertical shoots upward and masses of fibrous roots downward along their length. These shoots continually produce new buds that grow upward, adding more leaves and new horizontal rhizomes. This method of growth prevents complete burial by blown sand that accumulates around the base of the leaves, and plants have been known to outgrow sand accumulation rates of more than one meter per year. As the shoots elongate, they also branch, eventually producing the dense tussocks characteristic of old, healthy individuals. These dense stands may crowd out native species. This growth pattern, however, ideal on the shifting sands of the foredune, severely limits beach grass growth on the backdune. The shoots keep elongating even though they are not buried and, on the more stable backdune, the plants actually grow up out of the sand and are easily toppled and destroyed by wind and animals.

Distribution: *A. arenaria*, Oregon border to Los Angeles.

Flymus (grain) Rye Grass, Wild Rye

American dune grass or sea lime grass (*Elymus mollis*, pls. 11e, 12a) has an enormous range, extending from northern Alaska and around the entire Arctic south along the American coast to Point Conception. Since the introduction of marram grass (p. 186) to this coast in the late nineteenth century, *Elymus* has undergone a reduction in its importance because marram grass aggressively crowds it out. But several northern California beaches are still dominated by *Elymus*. It produces broad, blue-green shiny leaves from a thick rhizome. Its leaves are shorter than those of marram grass, but it is capable of keeping above the shifting sand. It does not form dense stands, however, and is usually accompanied by several other species interspersed among the clumps of leaves.

Distribution: *E. mollis*, Oregon border to Point Conception.

9

Synopsis of Common California Seashore Plants

The following list summarizes the phyla (divisions), classes, orders (algae only), families, genera, and species of algae and flowering plants discussed in this book. To aid in identification, some algae from different orders and families have been grouped together in the text because they have a similar external form. The list gives the correct taxonomic arrangement and can also serve as a checklist for those interested in keeping track of the plants they have encountered. Plants preceded by an asterisk are mentioned in the text but are not found in California. A listing of the characteristics of the various taxonomic levels is beyond the scope of this book but can be found in the works by Abbott and Hollenberg (algae) and Munz and Keck (flowering plants) given in the bibliography.

Algae

Cyanophyta (phylum: blue-green algae)
Chrysophyta (phylum: yellow-brown algae)
    Xanthophyceae (class)
        Vaucheriales (order)
            Vaucheriaceae (family)
                Vaucheria (genus)
                    longicaulis (species)

Chlorophyta (green algae)
   Chlorophyceae
      Ulotrichales
         Ulotrichaceae
            Ulothrix
               pseudoflacca
         Chaetophoraceae
            Trentepohlia
         Monostromataceae
            Monostroma
         Ulvaceae
            Blidingia
               minima
            Enteromorpha
               compressa
               intestinalis
            Ulva
               taeniata
      Prasiolales
         Prasiolaceae
            Prasiola
               meridionalis
      Cladophorales
         Cladophoraceae
            Chaetomorpha
               linum
               spiralis
            Cladophora
               columbiana
               graminea

Rhizoclonium
Spongomorpha
  coalita
Urospora
Codiales
  Bryopsidaceae
    Bryopsis
      corticulans
  Derbesiaceae
    Derbesia
      marina (including "*Hali-cystis ovalis*")
  Codiaceae
    Codium
      fragile
      fragile ssp. *tomentosoides*
      *magnum
      setchellii

Phaeophyta (brown algae)
  Phaeophyceae
    Ectocarpales
      Ectocarpaceae
        Ectocarpus
        Giffordia
          granulosa
    Chordariales
      Ralfsiaceae
        Ralfsia
      Corynophlaeaceae
        Cylindrocarpus
          rugosus

       Leathesia
          difformis
          nana
    Chordariaceae
       Analipus
          japonicus
Dictyosiphonales
    Dictyosiphonaceae
       Coilodesme
          californica
          plana
          rigida
    Punctariaceae
       Phaeostrophion
          irregulare
       Soranthera
          ulvoidea
Scytosiphonales
    Scytosiphonaceae
       Colpomenia
          peregrina
          sinuosa
       Endarachne
          binghamiae
       Petalonia
          fascia
       Scytosiphon
          lomentaria
Dictyotales
    Dictyotaceae
       Dictyopteris

        undulata

    Dictyota

        binghamiae

        flabellata

    Pachydictyon

        coriaceum

    Taonia

        lennebackeriae

    Zonaria

        farlowii

Desmarestiales

    Desmarestiaceae

        Desmarestia

            kurilensis

            latifrons

            ligulata var. *ligulata*

            ligulata var. *firma*

Laminariales

    Laminariaceae

        Agarum

            fimbriatum

        Costaria

            costata

        Hedophyllum

            sessile

        Laminaria

            dentigera

            ephemera

            farlowii

            sinclairii

        Pleurophycus

gardneri

Alariaceae
Alaria
marginata
Egregia
menziesii
Eisenia
arborea
Pterygophora
californica

Lessoniaceae
Dictyoneuropsis
reticulata
Dictyoneurum
californicum
Lessoniopsis
littoralis
Macrocystis
integrifolia
pyrifera
Nereocystis
luetkeana
Pelagophycus
porra
Postelsia
palmaeformis

Fucales
Fucaceae
Fucus
distichus
Hesperophycus

        harveyanus

     Pelvetia

       fastigiata

     Pelvetiopsis

       limitata

   Cystoseiraceae

     Cystoseira

       osmundacea

     Halidrys

       dioica

   Sargassaceae

     Sargassum

       agardhianum

       muticum

       palmeri

Rhodophyta (red algae)

  Bangiophyceae

    Goniotrichales

      Goniotrichaceae

        Goniotrichum

    Bangiales

      Erythropeltidaceae

        Smithora

         naiadum

        Erythrotrichia

      Bangiaceae

        Bangia

         fusco-purpurea

        Porphyra

         lanceolata

         nereocystis

perforata

Florideophyceae
Nemaliales
Acrochaetiaceae
Acrochaetium
desmarestiae
Rhodochorton
purpureum
Nemaliaceae
Nemalion
helminthoides
Helminthocladiaceae
Cumagloia
andersonii
Gelidiaceae
Gelidium
coulteri
nudifrons
purpurascens
pusillum
robustum
Pterocladia
capillacea
Cryptonemiales
Weeksiaceae
Constantinea
simplex
Peyssonneliaceae
Peyssonnelia
meridionalis
Hildenbrandiaceae

     Hildenbrandia
Corallinaceae
     Amphiroa
          zonata
     Bossiella
          californica ssp. *californica*
          californica ssp. *schmittii*
          orbigniana ssp. *dichotoma*
     Calliarthron
          cheilosporioides
          tuberculosum
     Corallina
          officinalis var. *chilensis*
          vancouveriensis
     Haliptylon
     Jania
          adhaerens
          crassa
          tenella
     Lithophyllum
          imitans
     Lithothamnium
          californicum
     Lithothrix
          aspergillum
     Melobesia
          marginata
          mediocris
     Mesophyllum
          conchatum
          lamellatum

Pseudolithophyllum
    neofarlowii
Serraticardia
Endocladiaceae
    Endocladia
        muricata
Cryptonemiaceae
    Grateloupia
        doryphora
    Halymenia
        californica
    Prionitis
        lanceolata
        lyallii
Kallymeniaceae
    Callophyllis
        flabellulata
        pinnata
        violacea
    Erythrophyllum
        delesserioides
Gigartinales
Nemastomataceae
    Schizymenia
        pacifica
Solieriaceae
    *Eucheuma
    Neoagardhiella
        gaudichaudii
    Opuntiella
        californica

Hypneaceae
    Hypnea
        valentiae var. *valentiae*
Plocamiaceae
    Plocamium
        cartilagineum
        violaceum
Gracilariaceae
    Gracilaria
        andersonii
        sjoestedtii
        textorii var. *cunninghamii*
        verrucosa
Phyllophoraceae
    Gymnogongrus
        leptophyllus
        linearis
        platyphyllus
    Stenogramme
        interrupta
Gigartinaceae
    *Chondrus
        crispus
    Gigartina
        agardhii
        canaliculata
        corymbifera
        exasperata
        harveyana
        papillata (including "*Pe-trocelis middendorffii*")

          spinosa

          volans

       Iridaea

          cordata

          flaccida

          heterocarpa

          lineare

       Rhodoglossum

          affine

          californicum

Rhodymeniales

  Rhodymeniaceae

     Botryocladia

        pseudodichotoma

     Fauchea

        laciniata

     Fryeella

        gardneri

     Maripelta

        rotata

     Rhodymenia

        californica

        pacifica

     Sciadophycus

        stellatus

  Champiaceae

    Coeloseira

    Gastroclonium

       coulteri

  Palmariaceae

    Halosaccion

        glandiforme
     Palmaria
        palmata var. *mollis*
Ceramiales
  Ceramiaceae
     Antithamnion
     Callithamnion
       pikeanum
     Centroceras
       clavulatum
     Ceramium
       codicola
       pacificum
     Microcladia
       borealis
       coulteri
     Neoptilota
     Platythamnion
     Ptilota
       filicina
     Ptilothamnionopsis
       lejolisea
     Tiffaniella
       snyderiae
  Delesseriaceae
     Acrosorium
       uncinatum
     Anisocladella
       pacifica
     Botryoglossum
       farlowianum

ruprechtianum
Cryptopleura
   corallinara
   lobulifera
   violacea
Gonimophyllum
   skottsbergii
Hymenena
   cuneifolia
   flabelligera
   multiloba
Myriogramme
   spectabilis
Nienburgia
   andersoniana
Phycodrys
   setchellii
Polyneura
   latissima
Dasyaceae
   Rhodoptilum
      plumosum
Rhodomelaceae
   Chondria
      californica
      nidifica
   Erythrocystis
      saccata
   Janczewskia
      lappacea
   Laurencia

pacifica
spectabilis
subopposita
Murrayellopsis
dawsonii
Odonthalia
floccosa
Polysiphonia
hendryi
pacifica
paniculata
Pterochondria
woodii
Pterosiphonia
haileyi
dendroidea
Rhodomela
larix

Flowering Plants

Anthophyta (flowering plants)
Monocotyledoneae
Jucaginaceae
Triglochin
maritimum
Zosteraceae
Phyllospadix
scouleri
torreyi
Zostera

     marina

  Juncaceae

   Juncus

    acutus

  Cyperaceae

   Scirpus

    acutus

    californicus

  Gramineae

   Ammophila

    arenaria

   Distichlis

    spicata

   Elymus

    mollis

   Monanthochloe

    littoralis

   Spartina

    * *alterniflora*

    foliosa

Dicotyledoneae

  Frankeniaceae

   Frankenia

    grandifolia

    palmeri

  Cruciferae

   Cakile

    edentula

    maritima

  Aizoaceae

   Mesembryanthemum

chilense

crystallinum

edule

Tetragonia

tetragonioides

Chenopodiaccae

Atriplex

leucophylla

Salicornia

bigelovii

virginica

Suaeda

californica

Nyctaginaceae

Abronia

latifolia

maritima

umbellata

Batidaceae

Batis

maritima

Plumbaginaceac

Limonium

californicum

Convolvulaceae

Calystegia

soldanella

Cressa

truxillensis

Cuscutaceae

Cuscuta

                    salina
        Leguminoseae
            Lathyrus
                    littoralis
        Onagraceae
            Camissonia
                    cheiranthifolia
        Compositae
            Ambrosia
                    chamissonis
            Jaumea
                    carnosa

Glossary

Acellular: not divided into cells.

Acrid: a sharp, bitter odor.

Aerenchyma: spongy plant tissue that facilitates gas exchange.

Agar: a colloidal substance found in the cell walls and intercellular spaces of some red algae.

Alga: a nonvascular plant in which all cells of the reproductive structures are fertile (pl, algae; adj, algal).

Alginate: a salt of alginic acid; a colloidal substance found in the walls of some brown algae.

Algology: the study of algae.

Anaerobic: occurring in the absence of oxygen.

Anastomosing: joined or united like the parts of a network.

Apex: the uppermost point; tip (pl, apices)

Articulated: having flexible joints between hard parts as in articulated coralline algae.

Assemblage: a collection of things; group.

Axial: on or along an axis.

Axis: a line with respect to which the thallus or branch is symmetrical.

Bacteria: a group of microscopic, generally nonphotosynthetic organisms with cell walls but lacking a membrane-bound nucleus.

Benthic: occurring on or related to the bottom of the ocean.

Blade: the more-or-less broad, flattened, foliose part of an erect alga.

Blue-green algae: generally small algae that lack a membrane-bound nucleus and have characteristic pigments giving them a blue-green, black, or red color.

Brackish: somewhat salty but not as saline as open ocean water.

Branch: a subdivision of the main body of a plant arising from an axis.

Branchlet: a small branch; often the last branch in a branch system.

Bryozoans: a group of aquatic invertebrate animals that bud to form erect or mosslike colonies.

Calcareous: containing calcium carbonate (like limestone).

Carpospore: a spore produced in the cystocarp which arises after fertilization on the female gametophytes of red algae; usually germinates to produce the diploid tetrasporophyte.

Carrageenan: a colloidal substance found outside the cells of some red algae.

Chlorophyll: the green photosynthetic pigment found in plants.

Chloroplast: a membrane-bound structure containing the photosynthetic system.

Compressed: somewhat flattened so that the cross section is elliptical.

Conceptacle: a cavity that opens to the thallus surface and contains reproductive structures (as in *Fucus*).

Corallines: the common name for a family of calcareous red algae.

Cortex: the outermost layer of cells or tissue in a thallus.

Cortical cells: cells in the cortex.

Corticated: having a cortex.

Crustaceans: a group of aquatic arthropods with a hard exoskeleton (crabs, barnacles, shrimp, etc.).

Crustose: in the form of a crust.

Cross wall: a transverse wall separating cells or parts of a thallus.

Cystocarp: the "fruit" that develops after fertilization in the red algae; occurs on the female gametophyte and contains carpospores.

Decumbent: lying down but with the tip ascending.
Desiccation: the process of drying out.
Determinate: having limited growth.
Dichotomous: branching by forking in pairs.
Diploid: having twice the basic (haploid) number of chromosomes; the chromosome number of most sporophytes.
Discoid: resembling a disk.
Distichous: arranged in two rows on opposite sides of an axis.
Dorsal: related to or near the back.
Drift: plant material that has broken loose from the bottom and is moving with the water or deposited on the shore.

Elongate: stretched out, slender.
Endemic: restricted to a particular location.
Endophytic: living within a plant.
Entire: continuous; without divisions.
Epiphyte: a plant living on another plant.
Epiphytic: living on the surface of plants.
Erect: standing up.

Fern: a nonflowering vascular plant.
Filamentous: having a single row of cells; generally hairlike.
Flabellate: fan shaped.
Flagellum: a microscopic, whiplike structure whose beating moves a cell (pl, flagella).
Foliose: broad and flat.
Frond: a single and usually leaflike part of a thallus; in *Macrocystis* the stipe plus floats and blades.
Fucoid: related to or resembling rock weeds (like *Fucus*).
Fungi: plantlike organisms that are unable to make their own food (mushrooms, molds, etc.).

Gametophyte: a plant that produces gametes (usually haploid).

Geniculum: an uncalcified joint between the calcified segments of articulated corallines (pl, genicula).

Geniculate: abruptly bent.

Genus: a category of biological classification between family and species (pl, genera).

Globose: globular, spherical.

Habit: the gross aspect of a plant.

Halogenated: containing a halogen (fluorine, chlorine, etc.).

Halophyte: a plant that grows in salty soil.

Haploid: having a single complete set of chromosomes; the chromosome number of most gametophytes.

Haptera: the rootlike basal outgrowths that form the holdfast in some brown algae (s, hapteron).

Haustoria: outgrowths of parasitic plants which enter the host and absorb nutrients.

Heteromorphic: having a life history in which the phases differ in morphology.

Holdfast: the basal attachment organ of an alga.

Hydroids: colonial invertebrate animals forming generally white, turflike growths.

Hygroscopic: readily absorbing and retaining moisture.

Hyphae: the small, threadlike structures that make up the vegetative body of many fungi; elongate, colorless filaments in some algae.

Interference: the interaction of two light waves to produce fringes, colors, etc.

Intergeniculum: a calcified segment between the uncalcified joints of articulated corallines.

Internode: a segment of a jointed structure lying between areas where branches are produced.

Invertebrate: lacking a spinal column.

Involucre: whorls of bracts situated below flowers or fruits.

Isomorphic: having a life history in which the phases have the same morphology.

Kelp: large brown algae.

Lichen: a group of plants formed from the association of an alga and a fungus.

Life history: the vegetative and reproductive phases through which an organism may pass during its life.

Limpet: a mollusk with an open low conical shell.

Liverwort: generally prostrate lower plants related to mosses.

Longitudinal: running lengthwise.

Macro-: large; visible without magnification.

Medulla: the central tissue of a thallus.

Membranous: thin, pliable, and often somewhat transparent.

Micro-: small; not visible without magnification.

Midrib: the thickened longitudinal axis of a thallus or blade.

Morphology: the form and structure of an organism.

Moss: a lower plant forming turflike mats; lacking flowers but reproductively different from algae.

Multinucleate: each cell containing more than one nucleus.

Mycorrhiza: a symbiotic association of a fungus with the roots of a higher plant (pl, mycorrhizae).

Nemathecium: a wartlike elevation containing reproductive structures.

Nitrogen fixation: the conversion of atmospheric nitrogen gas into nitrogen-containing compounds.

Node: the regions of an axis where branches arise

Obligate: restricted to one particular mode of life.

Palmate: resembling a hand with fingers spread.

Papilla: a short, nipplelike outgrowth (pl, papillae).

Parenchymatous: tissue composed of thin-walled cells produced from divisions in more than one plane.

Pectinate: with lateral branches arranged like the teeth of a comb.

Peltate: a circular blade with the stipe attached in the center of the lower surface.

Percurrent: extending through the entire length of a structure.

Pericentral cells: cells surrounding and derived from central axial cells.

Periwinkle: a type of snail.

Petal: one of the flower parts, usually conspicuously colored.

Phloem: the food-conducting tissue of vascular plants.

Photoperiod: the relative lengths of light and dark periods.

Phototropic: capable of directional orientation to light.

Phycobilins: reddish and bluish photosynthetic pigments.

Phycology: the study of algae.

Phylum: one of the primary taxonomic groups of organisms; between kingdom and class (pl, phyla).

Phytoplankton: floating plant life.

Pinnate: featherlike.

Planktonic: free floating.

Plastid: a membrane-bound intracellular structure containing pigments.

Polysiphonous: composed of lateral (pericentral) cells surrounding a central axis (siphon); as in *Polysiphonia*.

Polysporangia: sporangia producing more than four spores.

Polystichous: arranged in many ranks.

Proliferation: a vegetative outgrowth from a normally terminal structure; usually smaller than the structure that produces it.

Prostrate: lying along the substratum.

Radial: developing uniformly around a central axis.

Receptacle: the terminal portion of a branch bearing numerous conceptacles (as in *Fucus*).

Red tide: seawater discolored by the presence of large numbers of dinoflagellates (small, motile algae).

Reticulate: forming a network.

Rhizoid: a usually colorless unicellular or filamentous attachment structure; generally microscopic.

Rhizome: an underground stem; in the algae a prostrate, thickened axis.

Saccate: saclike.

Salinity: a measure of the amount of dissolved salts in seawater.

Sea urchin: a group of invertebrates covered by a hard shell (test) and articulated spines.

Secondary metabolites: compounds produced by organisms that are not of direct importance to their own metabolism (chemical processes).

Sepal: one of the modified leaves that usually enclose the other flower parts.

Septate: provided with walls or partitions.

Septum: a wall or partition.

Sessile: attached; not free to move.

Siphonous: multinucleate and tubular.

Sorus: a group or cluster of reproductive structures (pl, sori).

Sp.: abbreviation for species (one).

Spadix: a kind of flower spike with a fleshy axis.

Spathe: a large bract enclosing a flower cluster (spadix).

Spermatium: the nonmotile male gamete in the red algae (pl, spermatia).

Sporangium: a structure in which spores are produced (pl, sporangia).

Spore: a unicellular, asexual reproductive structure.

Sporophyll: a blade that produces spores.

Sporophyte: a plant that produces spores; generally the diploid generation in algal life histories.

Spp.: abbreviation for species (more than one).

Spring tides: tides with a greater than average range.

Ssp.: abbreviation for subspecies.

Stamen: flower structure composed of an anther (pollen-bearing portion) and its supporting stalk.

Stipe: the stemlike, usually basal part of a thallus.

Stolon: a horizontal branch near the base of a plant that produces new plants; a runner.

Surge: the back-and-forth motion produced by waves.

Symbiotic: living together in close association.

Terminal: at the tip or apex.

Tetrasporangium: a sporangium in which four spores are formed in a definite manner (tetrahedral, zonate, etc.; see algal structure and reproduction section).

Tetraspore: a spore produced in a tetrasporangium.

Thallus: the plant body of an alga.

Tunicates: a group of generally saclike or globular marine animals; sea squirts.

Turgid: swollen; distended.

Ultimate: final; the last in a series.

Upwelling: the process by which water rises from a lower to a higher depth.

Utricle: an inflated portion of a tubular thallus.

Var.: abbreviation for variety.

Vegetative: not reproductive and not associated with reproductive cells.

Ventral: related to or near the front.

Vesicle: a small air bladder or float.

Zoospore: a motile spore.

Bibliography

Seaweeds

Abbott, I. A., and E. Y. Dawson. 1978. *How to Know the Seaweeds*, 2d ed. Dubuque: William C. Brown Company Publishers. 141 pages. An excellent guide for identifying seaweeds found along the shores of the United States.

Abbott, I. A., and G. J. Hollenberg. 1976. *Marine Algae of California*. Stanford: Stanford University Press. 827 pages. The complete technical reference on California seaweeds.

Bold, H. C., and M. J. Wynne. 1978. *Introduction to the Algae*. Englewood Cliffs: Prentice-Hall, Inc. 706 pages. A very complete text if you are interested in learning more about the biology of both marine and freshwater algae.

Chapman, V. J. 1970. *Seaweeds and Their Uses*. London: Methuen and Co. Ltd. 304 pages. A discussion of seaweed utilization throughout the world.

Madlener, J. C. 1977. *The Seavegetable Book*. New York: Clarkson N. Potter, Inc. 288 pages. The entire book is devoted to the use of seaweeds as food with many excellent recipes.

Rhoads, S. 1978. *Cooking with Sea Vegetables*. Brookline, Mass.: Autumn Press. 136 pages. Another seaweed cookbook with an excellent introduction to the uses of seaweeds.

Waaland, J. R. 1977. *Common Seaweeds of the Pacific Coast.* Seattle: Pacific Search Press. 120 pages. A guide emphasizing the natural history of seaweeds in Washington.

Flowering Plants

Barbour, M. G., and J. Major, eds. 1977. *Terrestrial Vegetation of California.* New York: John Wiley. 1002 pages. A thorough treatment of terrestrial plant communities with excellent chapters on marsh and dune plants.

Ferris, R. 1970. *Flowers of Point Reyes National Seashore.* Berkeley: University of California Press. 119 pages. Includes many of the dune plants in excellent black-and-white drawings and a color key.

Munz, P. A. 1961. *California Spring Wildflowers.* Berkeley: University of California Press. 122 pages. Includes many of the flowering plants found along the coast.

————. 1964. *Shore Wildflowers of California, Oregon, and Washington.* Berkeley: University of California Press. 122 pages. A guide to the flowering plants of coastal strands and marshes.

Munz, P. A., and D. D. Keck. 1973. *A California Flora and Supplement.* Berkeley: University of California Press. 1905 pages. The complete technical reference on California higher plants.

General

Barbour, M. G., R. B. Craig, F. R. Drysdale, and M. T. Ghiselin. 1973. *Coastal Ecology: Bodega Head.* Berkeley: University of California Press. 338

pages. A generally nontechnical treatment of the ecology of a coastal region in northern California.

Hedgpeth, J. W. 1962. *Introduction to Seashore Life of the San Francisco Bay Region and the Coast of Northern California.* Berkeley: University of California Press (Natural History Guide). 136 pages. Contains a good discussion of coastal oceanography and is very helpful in identifying the animals associated with intertidal seaweeds.

Hinton, S. 1969. *Seashore Life of Southern California.* Berkeley: University of California Press (Natural History Guide). 181 pages. The southern California equivalent of Hedgpeth (above).

North, W. J. 1976. *Underwater California.* Berkeley: University of California Press (Natural History Guide). 276 pages. A natural history guide to the subtidal marine life of California with a discussion of subtidal seaweeds.

Ricketts, E. F., J. Calvin, and J. W. Hedgpeth. 1968. *Between Pacific Tides,* 4th ed. Stanford: Stanford University Press. 614 pages. A semitechnical treatment of the natural history of the animals found in the Pacific Coast intertidal zone.

Index

Italicized page numbers indicate location in keys; bold face page numbers indicate location of generic and other major descriptions; color plates are indicated by p. number; figures are indicated by f. number.

Abalone, 108
Abronia, **181**, f. 60
　latifolia, 181
　maritima, 181, f. 63a
　umbellata, 181, f. 63b
Acellular, 54
Acrochaetium desmarestiae, 69, 98
Acrosiphonia. See *Spongomorpha*
Acrosorium uncinatum, 39, 46, **146**, f. 49b
Aerenchyma, 165, 170
Agar, 103, 122, 132
Agardh, J. G. and C. A., 121
Agarum fimbriatum, 86
Alaria marginata, 5, 33, 81, f. 22a
Algae, 2, 9. See *also* Seaweed
　Blue-Green, 2, 3, 56, 163, p. 2b, p. 11d
　Brown. See Phaeophyta
　fossil, 92
　Green. See Chlorophyta
　mats, p. 11d, f. 56
　Red. See Rhodophyta
Algin, 73, 75, 78, 85
Algologist, 2
Alkali Heath. See *Frankenia*
　Alkali Weed. See *Cressa*
Ambrosia chamissonis, **184**, f. 65

Ammophila arenaria, **186**, 187, p. 12b
Amphiroa zonata, 111
Analipus japonicus, 35, **63**, f. 16a
Anisocladella pacifica, 141
Antheridium, f. 9, f. 10
Antithamnion, 37, 137, 139, p. 10c
Asterionella, p. 1c
Atriplex leucophylla, **184**, f. 64

Bacteria, 3, 175
Baker, K. D., 100
Bamboo, 74
Bangia fusco-purpurea, 37, 97, p. 7b, f. 27b
Barley, 175
Batis maritima, **166**, f. 57c
Beach, 4, f. 60
　Burr. See *Ambrosia*
　Primrose. See *Camissonia*
Black Brant, 159
Blade, 20, f. 4
Blidingia minima, 52
Bossiella, 36, **112**, 113
　californica ssp. *californica*, 112, f. 32b
　ssp. *schmittii*, 112
　orbigniana ssp. *dichotoma*, 112, f. 32a

219

Botryocladia pseudodichotoma,
  41, **136**, p. 10b
Botryoglossum, 39, 145, **146**
  farlowianum, 146, f. 50
  ruprechtianum, 147
Branching, 21
  types, f. 5
Bryopsis corticulans, 30, **51**, 54,
  f. 11f
Bryozoan, 14, 15, 128
Bulrush. See *Scirpus*

Cactus, 124
Cakile, 154, **177**, 179, 185
  edentula, 177
  maritima, 177, p. 12a, p. 12e,
    f. 60, f. 61a
California Current, 4, f. 1
Calliarthron, 37, 109, **113**, f. 3
  cheilosporioides, 113, p. 10e,
    f. 32c
  tuberculosum, 114, p. 10f
Callithamnion pikeanum, 44,
  139, f. 46c
Callophyllis, 44, **120**, 128,
  f. 37b
  flabellulata, 120, f. 37c
  pinnata, 120
  violacea, 120, f. 37a
Calystegia soldanella, **183**,
  p. 12c, f. 60
Camissonia cheiranthifolia, **185**,
  f. 66
Carposporophyte, f. 10
Carpospore, 24, f. 10
Carrageenan, 130, 132, 135
Cell, gland, 117, f. 34e
  structure, 20
Centroceras clavulatum, 139,
  f. 45a
Ceramium, 41, **138**
  codicola, 139
  pacificum, 138, f. 45b, f. 45c
Chaetomorpha, 30, **49**
  linum, 49, f. 11b
  spiralis, 49
Chlorophyll, 95, 96

Chlorophyta, 18, 20, 21, 22, 28,
  29, 47, 95
  reproduction, 22, f. 8
Chondria, 45, **151**
  californica, 151
  nidifica, 152, f. 52b
Chondrus crispus, 132
Cladophora, 30, **49**, 50
  columbiana, 50, p. 2e, f. 11c,
    f. 11d
  graminea, 50
Codium, 30, **54**
  fragile, 54, 139, f. 13a
    ssp. *tomentosoides*, 55
  magnum, 54
  setchellii, 55, f. 13b
Coeloseira, 136
Coilodesme, 31, **61**
  californica, 61, 92, f. 15a
  plana, 62
  rigida, 62
Colpomenia, 31, **62**, p. 2e
  peregrina, 63, f. 15c
  sinuosa, 63
Competition, 11, 14, 135, 162,
  179, 187
Conceptacle, 36, f. 9b
"*Conchocelis*," 100
Conferva, 86
Constantinea, 5, 40, **103**
  simplex, 103, f. 29
  subulifera, 104
Copepod, 66
Corallina, 37, 106, **111**, p. 2c,
  f. 2
  officinalis var. *chilensis*, 111,
    f. 31d
  vancouveriensis, 111, p. 7d,
    f. 31e
Corallines, 36
  articulated, 108, 109, 111,
    112, 113
  encrusting or crustose, 56,
    106, 108, 109, p. 7a,
    f. 3
Corals, 112
Cortex, 21, f. 6

Costaria costata, *33*, **85**, f. 23d
Crab, 15
Cressa truxillensis, **171**, f. 59c
Crustaceans, 66, 114, 136
Cryptopleura, *39*, **143**
  corallinara, 145
  lobulifera, 145
  violacea, 145, f. 48c
Cumagloia andersonii, *45*, *64*,
  101, p. 6d
Cuscuta salina, 166, p. 11b,
  p. 11c
Cylindrocarpus rugosus, *31*, 60,
  f. 14c
Cystocarp, 96, f. 10, f. 34b
Cystoseira osmundacea, *32*, 61,
  86, 90, 92, 151, f. 2, f. 3,
  f. 25a, f. 25b

Dead Man's Fingers. See
  Codium
Delesseriaceae, 143
Derbesia marina, 51, 56, 57
Desmarestia, *34*, **69**
  kurilensis, 70
  latifrons, 69
  ligulata var. *firma*, 69
    var. *ligulata*, 69, f. 18
Diatoms, *59*, 140, 163, p. 1c,
  f. 56. See *also* Microalgae
  and Phytoplankton
Dictyoneuropsis reticulata, 80
Dictyoneurum californicum, *35*,
  80, f. 21b
Dictyopteris undulata, *34*, **68**,
  f. 17f
Dictyota, 68, f. 17c
  binghamiae, 66
  flabellata, 65
Dictyotales, 68
Dinoflagellate, p. 1d
Distichlis spicata, **170**, 171,
  f. 56, f. 58b
Dodder. See *Cuscuta*
Doty, M., 132
Dulse, 128
Dune plants, 4, 25, **174**, f. 60

Ectocarpen, 59
Ectocarpus, 58
Egregia menziesii, 14, *32*, *62*,
  75, 83, 85, p. 3e, f. 2,
  f. 3, f. 20
Eisenia arborea, 14, *35*, 81, **83**,
  p. 3c, f. 3, f. 22b
Elymus mollis, **187**, p. 11e,
  p. 12a, f. 60
Endarachne binghamiae, 65
Endocladia muricata, 29, **44**,
  114, 135, p. 3a, f. 2, f. 33
Enteromorpha, 7, 29, 47, **51**,
  163, p. 1e, f. 12c
  compressa, 52, f. 11d
  intestinalis, 52, 63
Erythrocystis saccata, 40, 154,
  155, f. 52a
Erythrophyllum delesserioides,
  39, **118**, f. 36
Erythrotrichia, 98
Estuary, 4, 7, 122, 157, 159
Eucheuma, 132

Farlow, W. G., 66
Fauchea laciniata, *44*, **132**,
  f. 43b
Ferns, 4, 90
Filamentous, 37
Fish, 14, 15, 123, 148, p. 7c
Flowering plants, 4, 157, 162,
  174
  food uses, 159, 165, 173, 175,
  179, 182
  other uses, 166, 173, 181, 182
Fly larvae, 114
Frankenia, **170**
  grandifolia, 170, f. 56, f. 58d
  palmeri, 171
Fryeella gardneri, **44**, **132**, 141
Fucus distichus, 22, 28, *32*, **86**,
  88, 90, p. 5b, f. 2, f. 9b
Fungi, 4, 52, 175

Gametes, 2, 22
Gametophyte, 19, 22, p. 4b, f. 8,
  f. 9a, f. 10

Gardner, N., 141
Garibaldi, 148, p. 7c
Gastroclonium coulteri, 28, 41,
135, p. 9d
Gelidium, 43, 101, 122, f. 3
coulteri, 102, f. 28a
nudifrons, 103
purpurascens, 102, f. 28b
pusillum, 102
robustum, 103, p. 9c
Genicula, 111
Giffordia granulosa, 3, 30, 58,
f. 14a
Gigartina, 40, 41, 43, 46, 50,
128, 139, p. 2c, p. 3a,
p. 8a
agardhii, 105, f. 41b
canaliculata, 130, 132, p. 9a
corymbifera, 130, p. 8c
exasperata, 130, f. 2, f. 42a
harveyana, 130, p. 8d
leptorhynchos, 130, f. 41a
papillata, 29, 105, 114, 129,
135, p. 7e, f. 2
spinosa, 130, f. 42b
volans, 130
Glasswort. See *Salicornia*
Gonimoblast, f. 10
Gonimophyllum skottsbergii,
147
Goniotrichum, 98
Gracilaria, 5, 7, 42, 45, 120,
121, 152, f. 10
andersonii, 122
sjoestedtii, 122, p. 2a, f. 38b
textorii var. *cunninghamii*,
122, f. 38c
verrucosa, 122, f. 56
Grass, Arrow. See *Triglochin*
Beach. See *Ammophila*
Cord. See *Spartina*
Eel. See *Zostera*
Marram. See *Ammophila*
Rye. See *Elymus*
Salt. See *Distichlis*
Salt Cedar. See
Monanthochloe

Salt Flat. See *Monanthochloe*
sea, 25, 157, 162
Surf. See *Phyllospadix*
Grateloupia doryphora, 40, 41,
46, 117, f. 34c, f. 34d
Grazing, 11, 14, 15, 69, 97,
108, 133, 134, 135, 154,
p. 7b
Gymnogongrus, 42, 55, 108, 125
leptophyllus, 126, f. 40a
linearis, 126
platyphyllus, 126, p. 9b

"*Halicystis ovalis*," 29, 56, 105,
f. 13c
Halidrys dioica, 31, 62, 92, f. 2,
f. 25c
Haliptylon, 111
Halogenated compounds, 154
Halophyte, 162, 179
Halosaccion glandiforme, 29,
40, 136, f. 44
Halymenia californica, 41, 115,
f. 34a, f. 34b
Hedophyllum sessile, 33, 80,
f. 21c
Herbarium paper, 17
Herring, 123
Hesperophycus harveyanus, 28,
34, 90, f. 24b
Hildenbrandia, 105
Holdfast, 16, 19, 20, f. 4
Hottentot Fig. See
Mesembryanthemum
Hydroids, 15, 128
Hymenena, 39, 145, 147
cuneifolia, 145
flabelligera, 145, f. 48a
multiloba, 145
Hypersaline, 7
Hypnea valentiae var. *valentiae*,
45, 125, f. 40d

Ice Plant. See *Mesembry-
anthemum*
Intertidal zonation, 9–11, p. 2b,
p. 2c, p. 3a, f. 2

Iridaea, 5, 40, 131, **133**, f. 2
  cordata, 133, f. 43a
  flaccida, 28, 133, p. 3a
  heterocarpa, 133
  linaere, 133
Irish Moss. See *Chondrus*

Janczwskia lappacea, 37, **152**,
    f. 52b
Jania, 5, 36, **109**
  adhaerens, 109, f. 31c
  crassa, 111
  tenella, 111
Jaumea carnosa, **166**, f. 57b
Juncus acutus, 173

Kelp, 2, 18, 19, 58, 84, 95
  beds, 84
  Bull. See *Nereocystis*
  drift, 15, 72, f. 60
  Elk. See *Pelagophycus*
  Feather Boa. See *Egregia*
  forests, 8, 13, 15, 58, 73, 84,
      112, 114, 128, p. 3d,
      f. 3
  fronds, 74
  Giant. See *Macrocystis*
  rafts, 15
Kombu. See *Laminaria*

Lacuna, 133
Lamarck, 112
Laminaria, 33, 81, **84**, 133, f. 3,
    f. 23
  dentigera, 84, p. 2d, p. 3b,
      f. 23a
  ephemera, 85
  farlowii, 84, f. 23b
  sinclairii, 55, 84, f. 23c
Laminariales, 85
Lathyrus littoralis, **179**, p. 12c,
    f. 60, f. 62a
Laurencia, 43, 45, 108, **152**,
    154
  pacifica, 154, 155, f. 52a
  spectabilis, 154, 177, f. 52c
  subopposita, 154

Laver. See *Porphyra*
Leathesia difformis, 63
  nana, 63
Legume, 179
Lessoniopsis littoralis, 5, 35, 77,
    78, p. 2d, f. 21a
Lichen, 4, p. 1b
Life history, 18, f. 8
Limestone, 108, 112
Limonium californicum, **168**,
    f. 58c
Limpets, 77, 97, 161, p. 6e,
    p. 7b, f. 2
Lined Chiton. See *Tonicella*
Linnaeus, 54, 112
Lithophyllum imitans, 106,
    f. 30e
Lithothamnion californicum,
    106, f. 30b
Lithothrix aspergillum, 36, **109**,
    f. 31b
Littorines, f. 2
Liverworts, 4
Lobster, 14, 161

Macroalgae, 2
Macrocystis, 8, 13, 14, 15, 32,
    72, 73, 78, 81, 83, 84,
    85, 136, 151, f. 3, f. 9a
  integrifolia, 75
  pyrifera, 74, 90, p. 3d, p. 4b,
      p. 4c
Maripelta rotata, 104
Marsh plants, 8, 25, **162**, f. 56
Marsh Rosemary. See *Limonium*
Medulla, 21, f. 6
Melobesia, 36, **108**, 161
  marginata, 108
  mediocris, 108, f. 30a
Mesembryanthemum, **181**, f. 60
  chilense, 182
  crystallinum, 182, f. 62b
  edule, 182, f. 62c
Mesophyllum, 36, **109**, 114
  conchatum, 109, f. 31a
  lamellatum, 109
Microalgae, 2, 56, 163, p.1c, p.1d

Microcladia, 42, 44, 124, **139**
   borealis, 139
   coulteri, 139, p. 8a
Monanthochloe littoralis, **171**,
   f. 59a
Monostroma, 54
Morning Glory. See *Calystegia*
Mosses, 4
Murrayellopsis dawsonii, 149
Mussels, 77, 115
Mycorrhizae, 175
Myriogramme spectabilis, 44,
   146, f. 49a

Nemalion helminthoides, 40,
   45, **101**, f. 27d
Nemathecia, 125
Neoagardhiella gaudichaudii,
   45, **120**, 122, 123, 152,
   f. 6a, f. 38a
Neoptilota, 140
Nereocystis luetkeana, 32, 70,
   78, 100, 151, p. 5a, f. 3
Nienburgia andersoniana, 38,
   141, **143**, f. 48b
Nori. See *Porphyra*
Notoacmea insessa, 77
   paleacea, 161
Nullipores, 112

Oar Weed. See *Laminaria*
Odonthalia floccosa, 42, 62,
   157, f. 53a
Oogonium, f. 9
Opuntiella californica, 41, **124**,
   f. 39

Pachydictyon coriaceum, 5, 35,
   65, 68, f. 17a, f. 17b
Palmaria palmata, 128
Parasite, 147, 152, 166
Pelagophycus porra, 32, 72,
   p. 5c, f. 3, f. 19
Pelvetia fastigiata, 28, 35, 86,
   88, p. 5b
Pelvetiopsis limitata, 28, 35, 88,
   f. 2, f. 24a

Periwinkles, 97, p. 7b
Petalonia fascia, 33, 60, 65,
   f. 16b
"*Petrocelis middendorffii*," 50,
   105, 129, f. 30d
Peyssonnellia meridionalis, 37,
   105, f. 30c
Phaeophyta, 18, 21, 22, 29, **30**,
   58, 95
   reproduction, 22, f. 9a, f. 9b
Phaeostrophion irregulare, 65
Phillipi, 112
Phycobilin, 95, 96, 155
Phycodrys setchellii, 39, **140**,
   143, f. 47a
Phycologist, 2
Phyllospadix, 8, 14, 27, 63, 65,
   90, 98, 108, 133, 141,
   143, **161**, p. 3a, p. 3b,
   f. 2, f. 3
   scouleri, 161, p. 11a
   torreyi, 161, f. 55
Phytoplankton, 3, p. 1c
Pickle Weed. See *Salicornia*
Pigments, 95, 96, 155
Plantules, 68
Platythamnion, 137
Pleurophycus gardneri, 86
Plocamium, 44, 46, **124**, 139,
   f. 3
   cartilagineum, 124, p. 8b
   violaceum, 124
Point Conception, 4, 5, f. 1
Pollution, 8, 15, 52, 74, 90, 128,
   161
Polyneura latissima, 39, **142**,
   143, f. 47b
Polysiphonia, 38, **149**, 151, 163
   hendryi, 149
   pacifica, 149, f. 51a
   paniculata, 149
Polysiphonous, 149
Porphyra, 40, 70, 98, **99**, p. 6b,
   p. 6c
   lanceolata, 100
   nereocystis, 100
   perforata, 99

Postelsia palmaeformis, 5, 32, 77, 78, 83, 139, p. 4a, p. 12d
Potash, 75
Prasiola meridionalis, 54, p. 3a
Prionitis, 43, 46, 117, 139
  lanceolata, 118, f. 35a
  lyallii, 118, f. 35b
Pseudolithophyllum neofarlowii, 36, 106, p. 6e
Pterochondria woodii, 38, 151, f. 51d
Pterocladia capillacea, 43, 101, 103, 145, f. 28c
Pterosiphonia, 38, 43, 151, p. 10d, f. 3
  baileyi, 151
  dendroidea, 148, 151, f. 51b
Pterygophora californica, 33, 81, f. 3, f. 4
Ptilota filicina, 42, 140, f. 46a, f. 46b
Ptilothamnionopsis lejolisea, 114

Ragweed. See *Ambrosia*
Ralfsia, 31, 60, f. 14b
Red tide, 2
Reproduction, 18, 25, f. 8. See also Chlorophyta, Phaeophyta, Rhodophyta
Rhizoclonium, 19
Rhodochorton purpureum, 37, 97, f. 27a
Rhodoglossum, 42, 130, 131
  affine, 29, 129, 131, f. 41c
  californicum, 132
Rhodomela larix, 45, 62, 155, 157, f. 53b
Rhodophyta, 18, 21, 22, 29, 36, 81, 95
  pigments, 95, 96
  reproduction, 10, 22, 96, 149, f. 10
Rhodoptilum plumosum, 140
Rhodymenia, 42, 126, 133, 143, f. 6b

  californica, 128, f. 40c
  pacifica, 128, p. 10a
Rock Weed. See *Fucus*

Salicornia, 165
  bigelovii, 165
  virginica, 165, p. 11b, p. 11c, f. 56
Salt Bush. See *Atriplex*
  glands, 165, 170
  marsh, 4, 162
  Wort. See *Batis*
Sand Verbena. See *Abronia*
Sargassum, 32, 92
  agardhianum, 93, f. 26c
  muticum, 94, f. 26a
  palmeri, 93, f. 26b
Schizymenia pacifica, 41, 117, f. 34e, f. 34f
Sciadophycus stellatus, 104
Scirpus, 171
  acutus, 173
  californicus, 173, f. 59b
Scytosiphon lomentaria, 31, 60, 63, p. 5d
Sea Blite. See *Suaeda*
  Bottle. See "*Halicystis*"
  Fig. See *Mesembryanthemum*
  Grapes. See *Botryocladia*
  hare, 154
  Lavender. See *Limonium*
  Lettuce. See *Ulva*
  Nipples. See *Halosaccion*
  Otter, 14, 70
  Palm. See *Postelsia*
  pen, 70
  Rocket. See *Cakile*
  Sacks. See *Halosaccion*
  Sargasso, 92
  slug, 56, 154
  Spinach. See *Tetragonia*
  turtles, 159
Seaweed, 2, 7, 13, 14, 18, 20
  collecting, 16
  identification, 25, 27
  pressing, 16, 124, 140
  food uses, 54, 64, 71, 77, 85,

97, 100, 123, 128, 132
other uses, 56, 71, 73, 75, 81,
85, 97, 103, 112, 122,
130, 132, 155
Sedge, 171
Seep Weed. See *Suaeda*
Serraticardia, 111
Setchell, W. A., 141
Silky Beach Pea. See *Lathyrus*
Siphonous, 51
Smith, G. M., 98
Smithora naiadum, 27, 28, 40,
98, 161, f. 27c
Snails, 114, 133. *See also*
Abalone, *Lacuna*,
Limpets, Littorines,
Periwinkles, *Tegula*,
Tonicella
Soranthera ulvoidea, 31, 62,
155, f. 15b
Sori, 20, 97, f. 4, f. 9a
Southern Sea Palm. See *Eisenia*
Spartina, 163, 170
alterniflora, 165
foliosa, p. 11b, p. 11d, f. 56
Spermatia, 22, f. 10
Spiny Rush. See *Juncus*
Sponge, 14, 50
Weed. See *Codium*
Spongomorpha coalita, 30, 50,
f. 11e
Sporangia, 21, f. 9a
Spores, 14, 18, 21, 22, 29, 96,
122, f.7, f. 8, f. 9a, f. 10,
f. 30d, f. 51c
Sporophyll, 20, f. 4
Sporophyte, 19, 22, 24, p. 4c,
f. 8, f. 9, f. 10
Stenogramme interrupta, 39, 42,
126, 128, f. 40b
Stipe, 20, f. 4
Stolons, 128
Suaeda californica, 168, f.
58a
Subtidal zone, 13–15, f. 3
Surge, 14
Symbiosis, 56, 175

Taonia lennebackerae, 35, 68,
f. 17e
Tegula brunnea, 105, f. 30c
Temperature, 5, 58, 74
Tetragonia tetragonioides, 179,
p. 12a, f. 61b
Tetraspores, 21, f. 7
Thallus, 19
Tides, 9, 11
Tiffaniella snyderiae, 37, 147,
f. 51c
Tomatoes, 175
Tonicella lineata, 108, p. 7a
Trentepohlia, 3, p. 1a
Triglochin maritimum, 168,
f. 57a
Tule. See *Scirpus*
Tunicate, 50
Turban snails. See *Tegula*

Ulothrix pseudoflacca, 30, 47,
f. 11a
Ulva, 28, 30, 47, 51, 52, 86,
p. 2a, p. 2b, p. 2e, p. 6a,
f. 8
taeniata, 54, f. 12a, f. 12b
Upwelling, 5
Urospora, 49

Vaucheria longicaulis, 51, 163

Water motion, 13, 14, 85
Wild Radish, 177
Rye. See *Elymus*
Worms, 15, 56, 114, 128, p. 7a,
f. 56

Zonaria farlowii, 35, 66, 68,
f. 17d
Zonation, 10, 13, 135, 162,
174, p. 2b, p. 2c, p. 2e,
p. 3a, f. 2, f. 3, f. 56,
f. 60
Zoospores, 21
Zostera marina, 98, 157, f. 54,
f. 56
var. *latifolia*, 159, f. 54
Zygote, 22, f. 8